Graziano Aretusi

SUPERVISED CLASSIFICATION OF THERMAL HIGH-RESOLUTION INFRARED IMAGES
A case study for the diagnosis of Raynaud's Phenomenon

2017

Teramo, Italy

opendata.stats@gmail.com

ISBN 978-1-326-91085-3

All rights reserved to the author. It is not allowed to reproduce or use in no way this book, or even a part thereof, without the express permission from the author.

© Copyright by Graziano Aretusi, 2017

To Isabella, Matteo and Leonardo.

Table of Contents

Table of Contents . . . v

Abstract . . . vii

Introduction . . . 1

1 The Raynaud's Phenomenon and the IR Imaging . . . 5
   1.1 The Raynaud's Phenomenon . . . 5
   1.2 Infrared Imaging . . . 7
   1.3 The Data Collection . . . 8

2 Exploring Raynaud's Phenomenon by IR Imaging . . . 10
   2.1 The Re-warming Curves . . . 10
   2.2 Polar Representation . . . 12

3 Data Processing and Feature Detection in IR Imaging . . . 17
   3.1 Processing of Thermal High Resolution Infrared Images . . . 17
      3.1.1 Segmentation . . . 17
      3.1.2 Registration . . . 19
   3.2 Feature Extraction . . . 25
      3.2.1 Gaussian Markov Random Fields . . . 26
      3.2.2 Texture Analysis . . . 33
   3.3 The Dataset . . . 36
      3.3.1 The Dataset in the Spatial Domain . . . 36
      3.3.2 The Dataset in the Spatio-Temporal Domain . . . 40

4 Fisher's Linear Discriminant Analysis for the High Dimension/Small Sample Size Problems . . . 43
   4.1 Feature Subset Selection . . . 46
      4.1.1 Stepwise Subset Selection . . . 48
      4.1.2 Stepwise Subset Selection: Sensitivity Analisys . . . 50

		4.1.3 Improving Stepwise Subset Selection: the JSS+E Algorithm	52
4.2	Classifying Raynaud's Phenomenon		58
	4.2.1	FLDA in the Spatial Domain Approach	65
	4.2.2	FLDA in the Spatio-Temporal Domain Approach	72

Conclusions   **81**

Bibliography   **85**

Abstract

This study proposes a supervised classification approach for the differential diagnosis of Raynaud's Phenomenon, on the basis of functional infrared (IR) imaging data. The segmentation and registration of IR images are discussed and two texture analysis techniques are introduced to deal with the feature extraction problem, both in spatial and spatio-temporal frameworks. The classification of data from healthy subjects and from patients suffering from primary and secondary Raynaud's Phenomenon is performed by using Linear Discriminant Analysis on a large number of features extracted from the images. The feature selection problem are also considered and proposed a new subset selection algorithm, called JSS+E (Jackknifed Stepwise Selection with Exhaustive search), in order to improve the stepwise selection procedure. The results discussed for a dataset collected at ITAB laboratory in Chieti, allow to refine the experimental protocol in a completely new non-invasive way.

Introduction

Raynaud's Phenomenon (RP) is defined as an episodic vasoconstriction of small arteries and arterioles, in response to cold exposure or emotional stress. In particular, RP patients manifest high sensitivity to cold stimuli.

Typically, RP affects toes, the nose's tip and earlobes and it is particularly visible on the hand's fingers. RP patients share, in the initial history of their disease, clinical features and even morphological closeness in term of vascular and capillary involvement. The RP is distinguished in two forms. The primary form (also called Raynaud's Disease) and the secondary form (also called Raynaud's Syndrome). The differences between the two forms are that the former occurs only by itself and is not accompanied by other diseases; in contrast the secondary form occurs as a consequence of other diseases. The most common causes are connective tissue diseases and, in particular, scleroderma (in fact, 90% of people with scleroderma are affected by RP).

It is very important to distinguish between primary and secondary form because the latter can be characterised by irreversible tissue damage and can progress to necrosis or gangrene. Moreover, the Raynauds disease may precede the onset of secondary form[1] of several years, even before of any other haematic and immunological sign. Therefore an early diagnosis, with a proper classification of the RP, it is mandatory in order to establish the proper therapeutic strategy and to achieve the best prognosis.

Among the physiological measurement techniques currently in use to diagnose RP, there are *Nailfold capillary microscopy, cutaneous laser-Doppler flowmetry* and *Plethysmography* but none of these techniques are completely exhaustive; each of them, also, produces only a partial investigation, usually assessing only one finger for both hands.

---

[1] The primary RP can evolve into secondary RP in 12-13% of the cases.

Another useful tecnique to diagnose RP is the Infrared (IR) Imaging. Differently from the previous tecniques, IR imaging can process simultaneously multiple fingers, usually, subjecting patients to a cold stress to observe their rewarming capability. Thus, it is a low-invasive technique and allow to collect a temporal sequence of infrared images providing the map of the superficial temperatures of a given body by measuring the emitted infrared energy. In particular, O'Reilly [O'Reilly et al., 1992] inspected the temperature rewarming curves observing significant differences between healthy and unhealthy patient's patterns. Ammer [Ammer, 1996] uses the thermal gradient from the metacarpophalangeal joints to the finger tips to diagnose the RP; in particular, the study conducted on 71 patients suspected of suffering from RP, shows a sensitivity of 78,4% and a specificity of 72,4%. Schuhfried *et alii*[Schuhfried et al., 2000] propose a stepwise logistic regression approach classifying healthy patients with a sensitivity of 96% and a specificity of 62% and unhealthy patients with a sensitivity of 77% and a specificity of 73%. Anderson *et alii* [Anderson et al., 2007] use the distal dorsal difference at 30°C in a multinomial logistic regression approach; they classify the secondary form of the RP, within the unhealthy subject, with a sensitivity and a specificity both of 82%.

Remarking that it is known that the RP presents considerable intrasubject variability which must be considered, therefore in this work we propose a spatial approach which would allow to refine the experimental protocol in a completely non-invasive manner. Moreover, we propose a new spatio-temporal approach to the diagnosis of RP.

The work is structured in the following way:

- **Chapter 1: The Raynaud's Phenomenon and the IR Imaging.**

    This section is devoted to briefly introduce the RP describing its pathological and physiological characteristics. Also, it will describe the experimental protocol and the technology of digital infrared thermography used for data collection.

- **Chapter 2: Exploring Raynaud's Phenomenon by IR Imaging.**

    Based on the analysis of a series of rewarming curves, this section presents an exploratory approach of the RP. Specifically, we propose a functional analysis of the data from which we observe that the patients are characterised by different spatial and spatio-temporal temperature structures.

- **Chapter 3: Data Processing and Feature Detection in IR Imaging.**

In the third chapter we explore the feasibility of diagnosing RP both in spatial [Aretusi et al., 2010b] and spatio-temporal [Aretusi et al., 2010a] domains, i.e., by considering only the first detected image or the overall temporal sequence collected during the experiment, respectively. The advantages to work in the spatial domain are multiple and consist not only in the opportunity of simplifying the experimental protocol (reducing cost and time of computing and data collection), but also in the opportunity to perform an experiment which is fully (completely) non-invasive for the patient. By considering the spatio-temporal domain, the protocol is better able to grasp the dynamic of the phenomenon in relation to the temperature recovery capability. Before discussing the classification problem, a key step was the image processing to yield images comparable among the subjects. In particular, the goal is to segment the image into regions of interest and perform an image spatial transformation with respect to a reference image. Since the nature of the thermographic images is different with respect to the images in the visible spectrum, the conventional segmentation algorithms are not always feasible when applied to infrared data, so this make the processing step particularly difficult. Furthermore, since the images are not-stationary in space and in time, we assume that the temperature values are realizations of a Gauss Markov Random Field. For the classification purpose, along with the trend and spatio-temporal interaction parameters, some statistical information theory measures will be used to synthesise the temperature values correlation structure of a subject. In particular, cooccurrence matrices will be built which categorise the data in L levels.

- **Chapter 4: Fisher's Linear Discriminant Analysis for the High Dimension/Small Sample Size Problem.**

Finally, in this section the problem of variables selection will be addressed. Describing some standard selection techniques, such as the *stepwise selection*, a new subset selection algorithm will be proposed, called Jackknifed Stepwise Selection with Exaustive search (JSS+E), which is very useful in the case of large datasets (fat data), not only to select those features with significant relevance in terms of discriminatory power, but also to avoid the overfitting due to the presence of too many variables. The following table summarises the topics covered in this study.

Topics	Type				1D	2D	3D
	P	F	S	C			
Segmentation	X				X	X	X
Registration	X				X	X	X
GMRFs	X	X				X	X
Texture Analysis		X				X	X
FLDA				X		X	X
Stepwise DA			X	X		X	X
JSS+E DA			X	X		X	X

P=Processing
F=Feature Extraction
S=Feature Subset Selection
C=Classification

1D=Unidimensional (time)
2D=Bidimensional (space)
3D=Tridimensional (space-time)

Table 1: Topics covered in this study

Chapter 1

The Raynaud's Phenomenon and the IR Imaging

1.1 The Raynaud's Phenomenon

Raynaud's Phenomenon (RP) is a paroxysmal vasospastic disorder of small arteries, supplying the hands (Figure 1.1) and feet; also, it can occasionally affect the tongue, the nose, the ears, and the nipples.

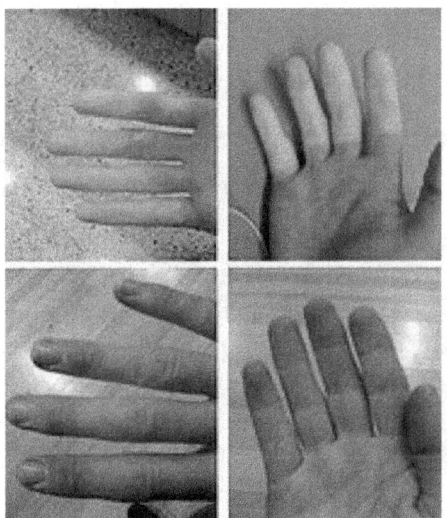

Figure 1.1: Examples of RP

RP produces an exaggerated response to the cold exposure and/or emotional stress [Belch, 2005] which results in a transient digital ischaemia. Typically, it is manifested by a tri-phasic discoloration of the extremities. There is an initial pallor due to vasospasm[1]. Peripheral cyanosis[2] due to oxygen deprivation may result and this is often followed by an hyperaemic phase[3] (Figure 1.2).

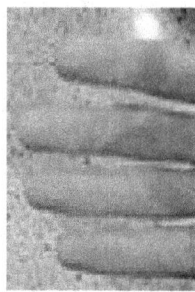 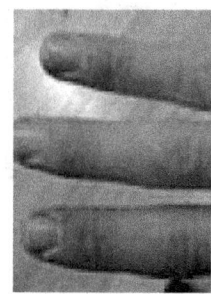 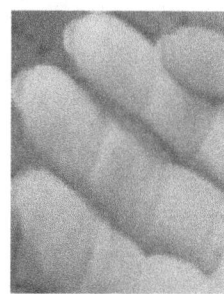

Figure 1.2: Tri-phasic discoloration of extremities: digital pallor (left), chianotic phase (middle), digital blush (right)

The presence of the initial pallor phase is mandatory for clinical diagnosis, but not all patients will experience the classic triad colour change, especially in mild cases. Raynaud is broadly classified into primary (PRP) and secondary phenomena. Primary RP is characterised by recurrence of vasospam without any underlying illness. Secondary RP is manifested by vasospam in the context of an associated pathology. Whilst secondary RP maybe precipitated by the use of certain drugs (for example beta blockers) and exposure to toxic agents [Block and Sequeira, 2001], the most common associations are connective tissue disorders such as systemic sclerosis. RP may precede the onset of the symptoms and signs of systemic sceloris by months or even years [Belch, 2005]. It is estimated that 12.6% of patients suffering from primary RP develops a secondary disease. Currently, the prevalence of primary RP varies greatly between populations, from 4.9%-20.1% in women to 3.8%-13.5% in men. The figures for secondary RP are not known since this depends on the underlying disease process; however, in general, the causes can increase morbidity and

---

[1] When exposed to cold temperatures, the blood supply to the fingertips is reduced and the skin turns pale or white (pallor) and becomes cold and numb.

[2] The skin colour turns blue due to the hemoglobin desaturation, that is scarcity of oxygen caused by the decreased blood flow.

[3] As a reaction to the rewarming process, blood returns to the district and the skin colour turns red and then back to normal. Often, this phase is accompanied by swelling and thingling. Symptoms are thought to be due to reactive hyperemias of the areas deprived of blood flow.

mortality of affected individuals. While PRP is generally characterised by an abnormal vasospastic response in absence of specific structural abnormalities, RP secondary to systemic sclerosis (SSc) is characterised by a peculiar re-arrangement of the microvascular structures [Kuryliszin-Moskal, 2005, Sato et al., 2005].

1.2 Infrared Imaging

Thermal infrared (IR) imaging is a technique introduced into the clinical context in the seventies and its use gave essentially modest results for many years. This was mainly due to the low spatial and thermal resolution of the available instruments and to the difficulty of adequately analyzing analogue images. In recent years, however, due to the progress of micro sensors and to the enormous improvement of contemporary computer systems for processing and analyzing data, a new generation of fully digital cameras, that offers excellent performance, was developed allowing both to make a powerful and accurate analysis of scanned images.

Thermal IR imaging has been widely used in medicine to evaluate cutaneous temperatures. IR imaging is a non-invasive technique providing the map of the superficial temperatures of a given body by measuring the infrared energy emitted [Merla et al., 2002b]. Since the cutaneous temperatures depend on the local blood perfusion and thermal tissue properties, IR imaging provides important indirect information on circulation, thermal properties and thermoregulatory functionality of the cutaneous tissue. Several IR imaging studies have been performed so far to attempt at differentiating primary from secondary RP, often in combination with the monitoring of the finger response to a controlled cold challenge. In fact, PRP, SSc and healthy controls (HCs) show different thermal recoveries in consequence of the same standardised functional stimulation [Clark et al., 1999, Di Carlo, 1995, Hahn et al., 1999, Herrick and Clark, 1998, Merla et al., 2002a].

The availability of new digital and high-resolution devices for thermal imaging of military origin opened new possibilities in the biomedical field, allowing a quantitative based diagnostic approach.

In particular, the camera used in this work (Figure 1.3) is a 14-bit digital thermal camera

(FLIR SC3000 QWIP, Sweden) sensitive in the 8-9 μm band with sensory array of 256×256 pixels, which corresponds to a spatial resolution of up to 0.14 mm (about 0.5 milliradians). The resolution in temperature is equally remarkable, reaching a value of 0.02°C.

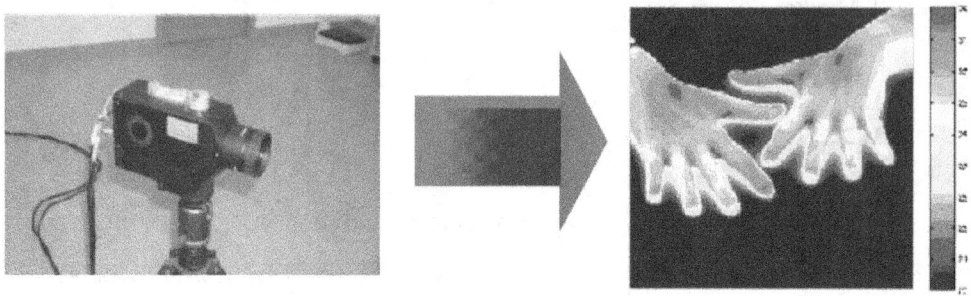

Figure 1.3: Examples of a digital thermal camera (left) and IR image (right)

Since the entire process of image acquisition is fully digital and remarkably fast[4], the system can be used to study both the spatial and spatio-temporal thermal properties of the process (e.g. monitoring the thermal recovery capability of a subject after an inducted functional test).

1.3 The Data Collection

Data for this study were provided by the Functional Infrared Imaging Lab - ITAB, Institute for Advanced Biomedical Technology, at the School of Medicine of the G. D'Annunzio University, Chieti, Italy. The study was approved by the local Institutional Review Boards and Ethics Committees. All the subjects gave their informed written consent prior to being enrolled for the experiment.

Data consist of temperature images documenting the thermal recovery from a standardised cold stress experiment on both hands [Merla et al., 2002a].

Specifically, the data refer to $K = 44$ subjects: 13 healthy controls (HCs), 14 Primary RP (PRP), and 17 Secondary RP (SSc). The patients were classified according to the American College of Rheumatology criteria and standard exclusion criteria were observed [Merla et al., 2002b]. Patients underwent thermal IR imaging after having observed standard

---

[4]The system is able to acquire more than 30 images per second.

preparatory rules to the test [Merla et al., 2002a] such as to abstain from smoking, caffeine and alcohol.

In particular, measurements were undertaken in a controlled environment with no direct ventilation, a temperature of about 23°C and a humidity of 50-60%. After an initial acclimatization phase of 20 minutes to minimise intra-subject variability reflecting the patient's sympathetic tone [Anderson et al., 2007], participants were seated with both hands placed on a table covered with a black non-reflective sheet and with forearms exposed. The experimental protocol expects a first stage of 2.5 minutes for the acquisition of the individual basal temperature; then a cold stress is induced by immersion of the hands in 3 liters of water at approximately 10°C for 2 minutes (while wearing thin plastic gloves). Finally, for the successive 20 minutes, the response to the cold stress is studied by measuring the rewarming process capability of each individual. For each subject, the spatio-temporal heating distribution is thus represented by a temporal sequence of 45 images. Figure 1.4 provides a pictorial representation of the functional test.

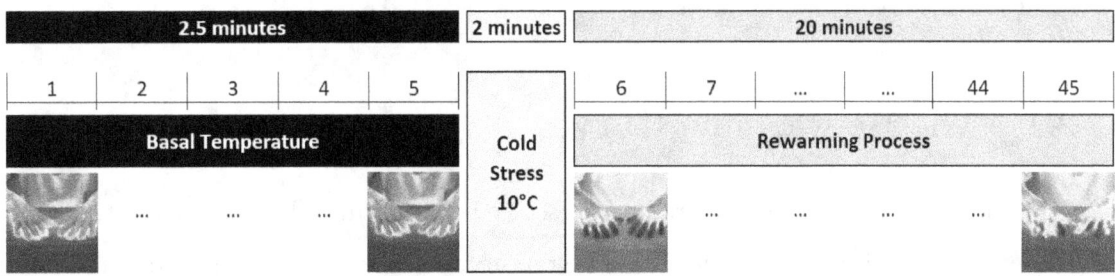

Figure 1.4: The experimental protocol

Thus, for the k-th subject ($k=1,2,\ldots,44$), the data matrix consists of a temporal sequence of 45 infrared images, each of dimension 256×256.

Chapter 2

Exploring Raynaud's Phenomenon by IR Imaging

To gain information about the behaviour of the RP with particular attention to its spatial and spatio-temporal temperature distributions, we present here an exploratory approach by IR imaging. In this section our goal is just to explore the groups (HCs, PRP, SSc) to identify some features of interest.

2.1 The Re-warming Curves

The aim of the experimental protocol described in the first chapter, is to study the re-warming capabilities of a subject as a response to a cold stress. Each individual may be represented by a rewarming curve summarizing his own temperature recovery process.

In particular, for each subject and at a specific time t, nail beds were considered as regions of interest from which to extract the temperature values average u_{hf} where $h = 1, 2$ represent left and right hand, respectively and $f = 1, 2, 3, 4, 5$ represent finger's order from the thumb to pinkie. These regions are the most sensitive to temperature changes and therefore are best placed to represent them. Thus, we may depict the objects (t, u_{hf}) on a coordinate system. For fixed values of h and f we obtain a rewarming curve[1] summarizing

---

[1] Although measured at discrete intervals, rewarming curves have an underlying continuous functional form and can thus be considered as samples of functional observations. The replication of curves, one for each finger, invites an exploration of the ways in which they vary so that for example, classical summary statistics would undoubtedly be helpful to explore their features. The basic idea of functional data analysis is that we consider a rewarming curve as single entity, rather than a sequence of observations. According to this

the temperature recovery process of the region of interest.

An example of rewarming curve extraction is depicted in Figure 2.1.

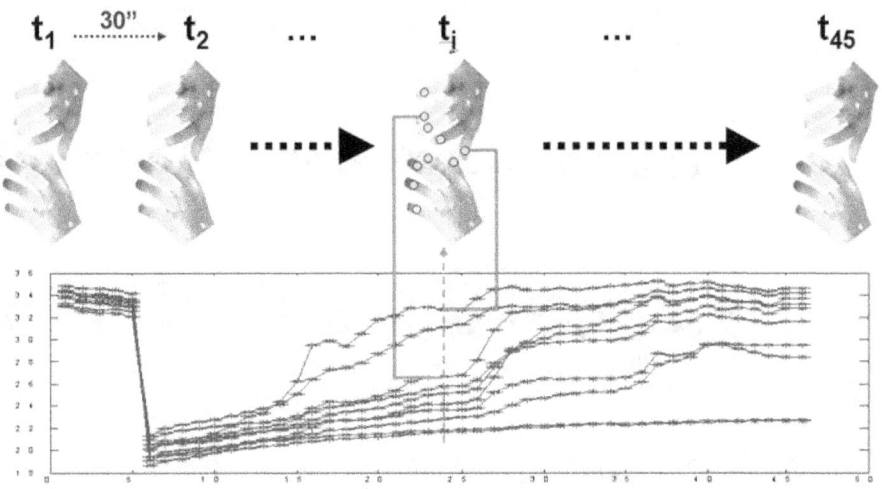

Figure 2.1: Example of rewarming curves extraction

The application of a standardised cold stress determines an immediate temperature drop from the stationary steady state. The recovery is monitored starting at the sixth image of the series, that is just after the cessation of the cold stress (see Figure 1.4). Figure 2.2 represents the rewarming curves for all the subjects.

At a first sight, the dynamics of the rewarming curves appear quite different in the three classes. The PRP group exhibits an extremely low interindividual variability, whereas the SSc group shows both between fingers and interindividual variability, according to the advancement and the presentation of the pathology [O'Reilly et al., 1992, Di Carlo, 1995, Block and Sequeira, 2001, Merla et al., 2002a]. Also, the HCs group recovers to the baseline temperature after a few minutes, while PRP patients do not appear to activate any effective rewarming capability. SSc patients appear to show an intermediate recovery behaviour. Therefore, the three groups appear to show different dynamics. This is particularly evident from the figure 2.3 which shows three typical examples (one for each group) of RP.

---

philosophy, the data set can be interpreted as a set of functions which can be manipulated in various ways. A key reference for FDA is the book by Ramsay and Silverman [Ramsay and Silverman, 1997]. Moreover, a reference for the study of RP in a functional data analysis approach is the work of Di Zio, Ippoliti and Merla [Di Zio et al., 2009].

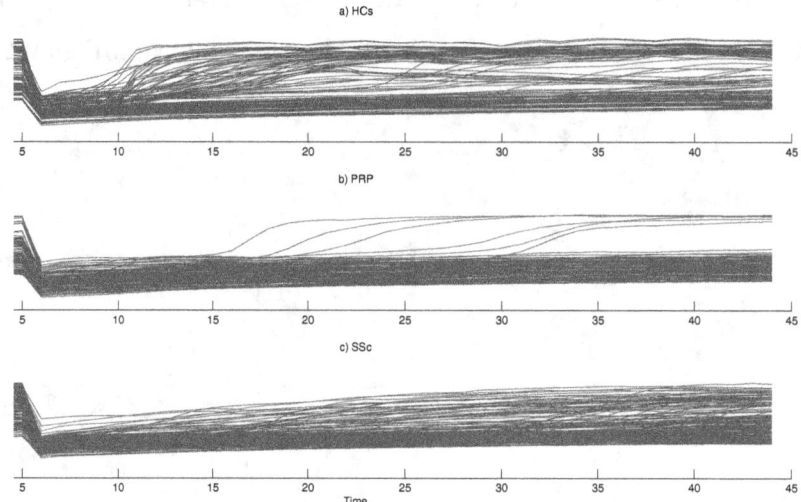

Figure 2.2: Re-warming curves for each group (HCs, PRP, SSc)

2.2 Polar Representation

Figure 2.2 shows that several rewarming curves do not reach specific temperature thresholds, such as the room temperature fixed at 23°C. At the same time, due to the different thermal and blood flow characteristics, it can also be observed that there are several fingers that are able to reach the same threshold temperature after different time points. Given the different pattern shown by healthy and unhealthy rewarming curves of the same subject, it appears extremely important to preserve the information of each single finger. To get more insight, we thus propose to represent the 10 fingers of a subject on a polar diagram with each finger represented by a radius placed at a fixed distance of 36° from each other. The radius, then, represents the time needed by the finger to reach a fixed temperature threshold, say τ. For a given threshold, we thus have a polar representation of a subject and, by joining the radii extremities, we obtain irregular planar figures. As an example, for $\tau = 21$°C, Figure 2.4 below shows a comparison of polar representations for three typical subjects belonging to HCs, PRP and SSc groups, respectively.

As can be seen, the healthy control seems to present a more symmetrical behaviour between the hands with respect to the PRP and SSc. Moreover, while the healthy control

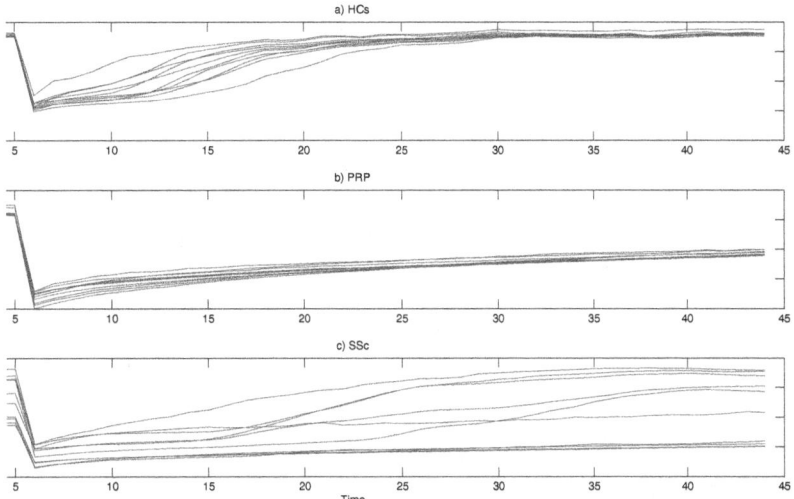

Figure 2.3: Typical examples of RP rewarming curves (HCs, PRP, SSc)

and PRP show almost a regular pattern, the PRP reaches the temperature threshold of 21°C more slowly. The SSc individual is much more irregular, with some of the fingers not reaching the fixed temperature threshold at all. Notice that for illustrative purposes, the fingers that do not reach the temperature threshold within the end of the experiment (marked black lines), are conventionally represented as a point at the origin (Figure 2.4c). Furthermore, in the representation is also imposed a condition of circularity in which the radii represent the fingers according to their natural order distinguishing by left (L) and right (R) hands, respectively.

In previous polar representations, was considered a fixed temperature threshold; in particular we set $\tau = 21$°C because it is interesting to observe the behaviour close to the room temperature. It is also notable to observe the behaviour of the subjects for different levels of τ. So, considering the interval $\tau \in [15; 35]$, it is possible to build a tri-dimensional (3D) polar representation of the rewarming capability of the subject for different temperature thresholds. Figure 2.5 shows 3D polar representations for typical RP patients.

For a fixed level of τ, we have the same polar representation as before and the same interpretation for the diagram. Moreover, joining all the points for each level of τ, it produces a solid figure that represent the behaviour of the subject after the cold stress.

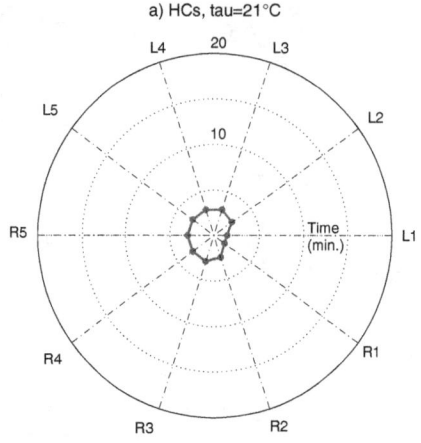
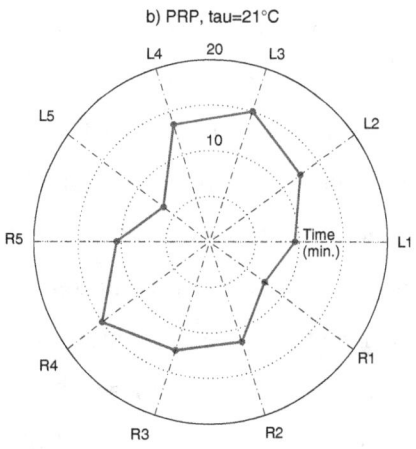
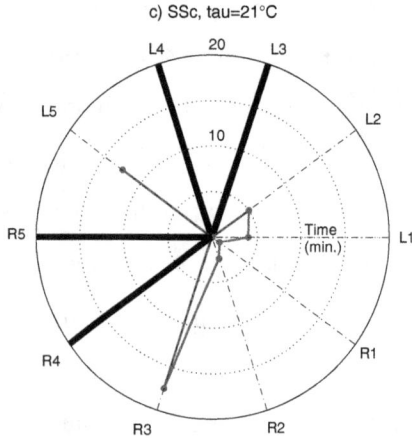

Figure 2.4: Polar representation for three typical subjects belonging to HCs (a), PRP (b) and SSc (c) groups. Each radius represents the time needed by each finger to reach the fixed threshold temperature of $\tau = 21°C$

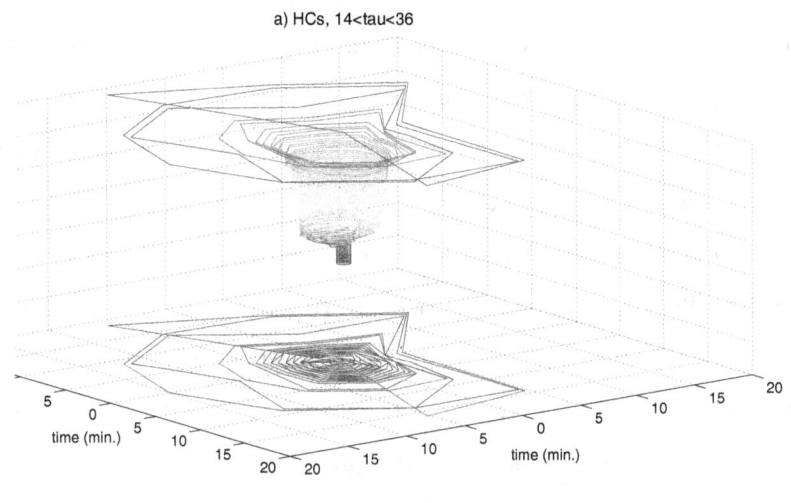
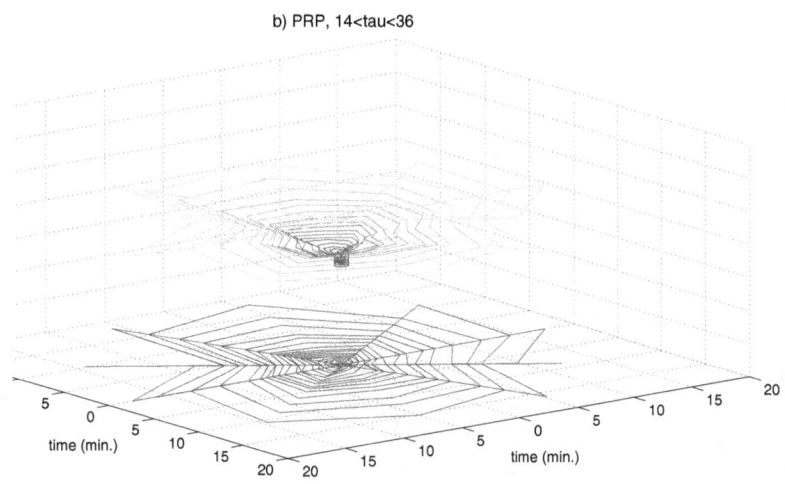
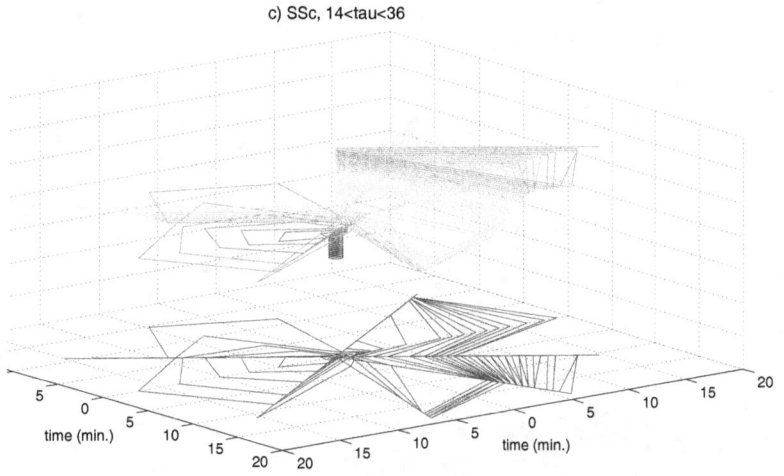

Figure 2.5: 3D Polar representation for three typical subjects belonging to HCs (a), PRP (b) and SSc (c) groups

Indeed, while the healthy control is able to reach temperature levels even beyond the 23°C of the room, the PRP subject is able to reach the temperature only down the threshold 23°C. The SSc patient shows a very complex behaviour, presenting some fingers that are able to recover certain temperature thresholds beyond 23°C while other fingers appear unable even to recover low temperature levels.

The observation of the various graphical representations shows that groups are characterised by different dynamics in space and time, although such differences among the groups are not always so pronounced. In any case, however, it is enough to assume that groups are characterised by different spatial and spatio-temporal temperature structures. Therefore, we will work on it assessing its nature and size.

Chapter 3

Data Processing and Feature Detection in IR Imaging

3.1 Processing of Thermal High Resolution Infrared Images

It is often a necessary step to carry out a processing of the images before a desired quantitative analysis. In particular, prior to the feature extraction for the classification of RP patients, our aim is to construct contours in the image to partition it into regions of interest (segmentation) and to perform a spatial transformation with respect to an oriented reference image in order to compare images through the subjects (registration). Because of the inherent differences between infrared and visible phenomenology, a number of fundamental problems arise when trying to apply traditional processing methods to the infrared [Scribner et al., 2000, Maldague, 2001].

3.1.1 Segmentation

The nature of thermal images is quite different from that of a conventional intensity image. In general, the latter encodes several physical properties such as reflectance, illumination, material of object surface, etc, to form the shape-related data, while a thermal image is formed from the heat distribution of an object[1].

---

[1] Specifically, thermographic images depict the electromagnetic radiation of an object in the infrared range which is about 6-15 μm.

Therefore, it is obvious that the conventional segmentation algorithms may not be feasible when they are applied to a thermal image [Chang et al., 1997, Heriansyak and Abu-Bakar, 2009]. The purpose of thermal image segmentation is to separate objects of interest from its surroundings which is usually represented by warm, hot or more generally by some thermal features that present a certain uniformity. In such cases, it is possible to perform a segmentation by using a threshold procedure (Figure 3.1).

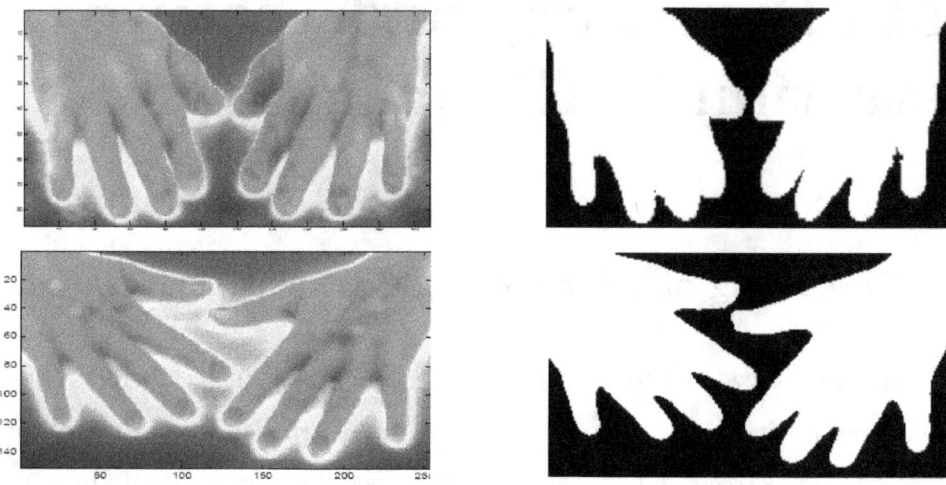

Figure 3.1: Examples of IR image segmentation using threshold procedure

However, due to the slight blurring caused by the infrared imaging process, it is possible to find that the boundary between a hand and the background is not that sharp for some images (see for example Figure 3.2).

In such cases, the images were segmented manually, pixel by pixel (Figure 3.3).

It is very often true that the infrared image acquisition is a part of the problem in the infrared segmentation process. In fact, the technology required for IR imaging is much less mature respect to that used in visible imaging [Scribner et al., 2000]. Moreover, it is known that the infrared images are often degraded by a number of reasons [Ibarra-Castanedo et al., 2004].

It is possible to avoid such problems in segmentation by means of the latest technology. In fact, nowadays, cameras are available that collect both visual and thermal images, so to perform a fusion between dedicated and nondedicated algorithms for object identification

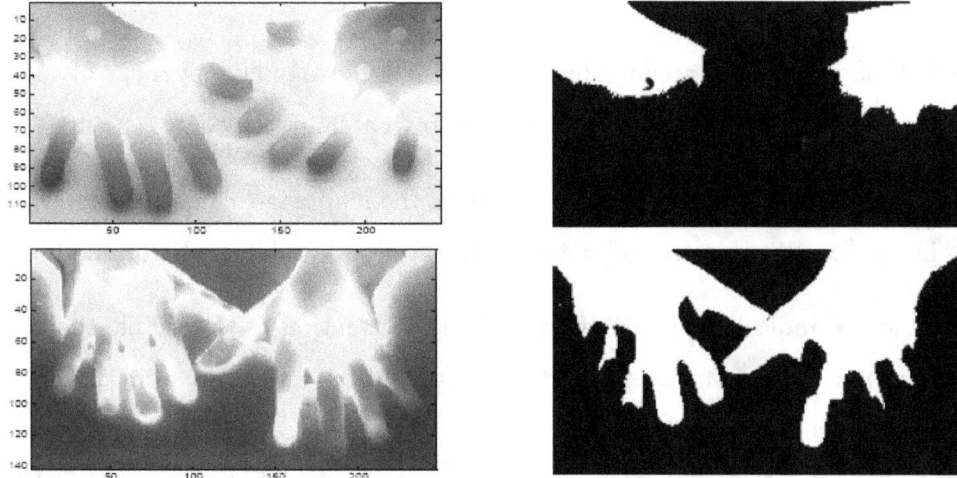

Figure 3.2: Examples of IR image segmentation using threshold procedure

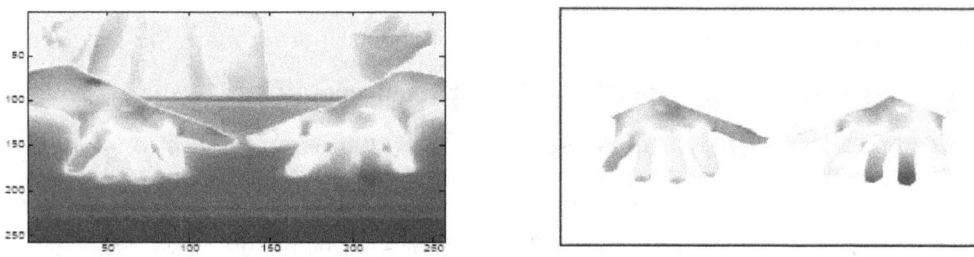

Figure 3.3: Examples of manual IR image segmentation

in infrared images (i.e. performing a segmentation on the visual image and then overlaying the segmented visual image on the thermal image).

3.1.2 Registration

Image registration is the process of geometrical alignment of two images, a sensed image with respect to a reference image, required to obtain more complete and comparable information through the subjects [Brown, 1992, Dryden, 1997]. The majority of registration methods consists of the following steps [Zitova and Flusser, 2003]:

- Manual or automated detection of distinctive objects represented by the so-called control points (input points on the sensed image and base points on the reference image).

- Estimation of the mapping function aligning the two images by mean of an established correspondence between the two images by matching the control points.

- Resample of the sensed image by mean of the mapping function (image values in non integer coordinates are computed by the appropriate interpolation method).

Due to the radically different ways of image formation in the visible spectrum and in thermographic images, many methods for registration of images work poorly or do not work at all [Jarc et al., 2007]. A reasonable way to practice is to detect the control points manually, usually by using an aided procedure, so to use a global mapping model to estimate a set of mapping function parameters valid for the entire image and, finally, to determine the registered image data from the sensed image using the target coordinates system of the reference image (base points) and the inverse of the estimated mapping function.

Let us denote $(x_i, y_i)'$ a point on the cartesian coordinates system xy of the sensed image A and $\mathbf{a}_i = (x_i, y_i, 1)'$ its homogeneous coordinates representation. Similarly, let us denote $(u_i, v_i)'$ a point on the cartesian coordinates system uv of the reference image B and $\mathbf{b}_i = (u_i, v_i, 1)'$ its homogeneous coordinates representation. For two-dimensional spaces, an affine transformation can be written as $B = HA$ where the map H (from xy to uv) preserves straight parallel lines and ratios of distances (shape) so that

$$H = \begin{bmatrix} h_{11} & h_{12} & h_{13} \\ h_{21} & h_{22} & h_{23} \\ 0 & 0 & 1 \end{bmatrix}$$

is a composition of the following maps H_T, H_M, H_R and H_S:

- $H_T = \begin{bmatrix} 0 & 0 & t_x \\ 0 & 0 & t_y \\ 0 & 0 & 1 \end{bmatrix}$ represent a translation,

- $H_M = \begin{bmatrix} m_x & 0 & 0 \\ 0 & m_y & 0 \\ 0 & 0 & 1 \end{bmatrix}$ represent a magnification,

- $H_R = \begin{bmatrix} cos(\theta) & -sin(\theta) & 0 \\ sin(\theta) & cos(\theta) & 0 \\ 0 & 0 & 1 \end{bmatrix}$ represent a rotation of θ degrees in an anticlockwise direction,

- $H_S = \begin{bmatrix} 1 & s_x & 0 \\ s_y & 1 & 0 \\ 0 & 0 & 1 \end{bmatrix}$ represent a shear.

Therefore, detecting n input points $P_A = (\mathbf{a}_0, \mathbf{a}_1, \ldots, \mathbf{a}_{n-1})$ corresponding to n base points $P_B = (\mathbf{b}_0, \mathbf{b}_1, \ldots, \mathbf{b}_{n-1})$, where $\mathbf{a}_i = (x_i, y_i, 1)'$ and $\mathbf{b}_i = (u_i, v_i, 1)'$, we obtain that the geometrical transformation of a point from A onto B is given by

$$P_B = HP_A.$$

The solution for H that provides the minimum mean-squared error is

$$H = P_B P_A^+$$

where $P_A^+ = P_A^T (P_A P_A^T)^{-1}$ is the right pseudo-inverse of P_A.

Since the distance and the angle between the thermal camera and the scene are not always the same with respect to the reference image, the perspective projection model [Brown, 1992, Zitova and Flusser, 2003] should be used to handle changes caused by a tilt of the image plane. The perspective projection transformation (aka planar homography) maps points in a tri-dimensional space to points on a plane (Figure 3.4).

So, by means of a transformation matrix H

$$H = \begin{bmatrix} h_{11} & h_{12} & h_{13} \\ h_{21} & h_{22} & h_{23} \\ h_{31} & h_{32} & h_{33} \end{bmatrix}$$

it is possible to map a point on A onto a point on B, considering that a planar homography maps points such as

$$u_i = \frac{h_{11}x_i + h_{12}y_i + h_{13}}{h_{31}x_i + h_{32}y_i + h_{33}}$$

and

$$v_i = \frac{h_{21}x_i + h_{22}y_i + h_{23}}{h_{31}x_i + h_{32}y_i + h_{33}}.$$

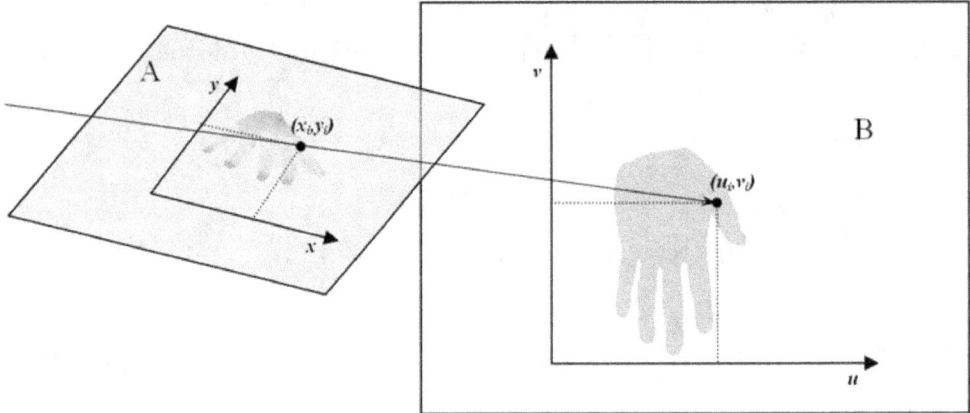

Figure 3.4: Prospective projection transformation

Thus, since the transformation matrix is defined up to a constant[2], it is possible to set $h_{33} = 1$. So, rearranging the equations as

$$u_i = h_{11}x_i + h_{12}y_i + h_{13} - h_{31}x_iu_i - h_{32}y_iu_i$$

and

$$v_i = h_{21}x_i + h_{22}y_i + h_{23} - h_{31}x_iv_i - h_{32}y_iv_i$$

we have that

$$\begin{bmatrix} u_0 \\ v_0 \\ u_1 \\ v_1 \\ \vdots \\ u_{n-1} \\ v_{n-1} \end{bmatrix} = \begin{bmatrix} x_0 & y_0 & 1 & 0 & 0 & 0 & -x_0u_0 & -y_0u_0 \\ 0 & 0 & 0 & x_0 & y_0 & 1 & -x_0v_0 & -y_0v_0 \\ x_1 & y_1 & 1 & 0 & 0 & 0 & -x_1u_1 & -y_1u_1 \\ 0 & 0 & 0 & x_1 & y_1 & 1 & -x_1v_1 & -y_1v_1 \\ \vdots & \vdots & \vdots & \vdots & \vdots & \vdots & \vdots & \vdots \\ x_{n-1} & y_{n-1} & 1 & 0 & 0 & 0 & -x_{n-1}u_{n-1} & -y_{n-1}u_{n-1} \\ 0 & 0 & 0 & x_{n-1} & y_{n-1} & 1 & -x_{n-1}v_{n-1} & -y_{n-1}v_{n-1} \end{bmatrix} \begin{bmatrix} h_{11} \\ h_{12} \\ h_{13} \\ h_{21} \\ h_{22} \\ h_{23} \\ h_{31} \\ h_{32} \end{bmatrix}$$

and the parameters $(h_{11}, h_{12}, h_{13}, h_{21}, h_{22}, h_{23}, h_{31}, h_{32})'$ can be found by multiplying both sides with the pseudo-inverse of the big matrix of coordinate terms.

---

[2] The transformation matrix H has 8 degrees of freedom.

It is important to note that the mapping function H can locate the points on the grid (integer coordinates) of A in non integer coordinates on B. So, it is necessary to interpolate the points on an integer coordinates system of the reference image B.

A less computational expensive way to forward [Zitova and Flusser, 2003], is to use the inverse of the calculated transformation matrix H so that $Q_A = H^{-1}Q_{Bg}$ where Q_{Bg} and Q_A represent, respectively, the known homogeneous grid coordinates of B and the projection from B to A. Therefore, it is possible to find the value at each point Q_A given from the values of Q_{Ag}, the homogeneous integer coordinates of A, by using the bilinear interpolation method [Semmlow, 2004, Pratt, 2007]. In this way neither holes nor overlaps can occur in the registered image (Figure 3.5).

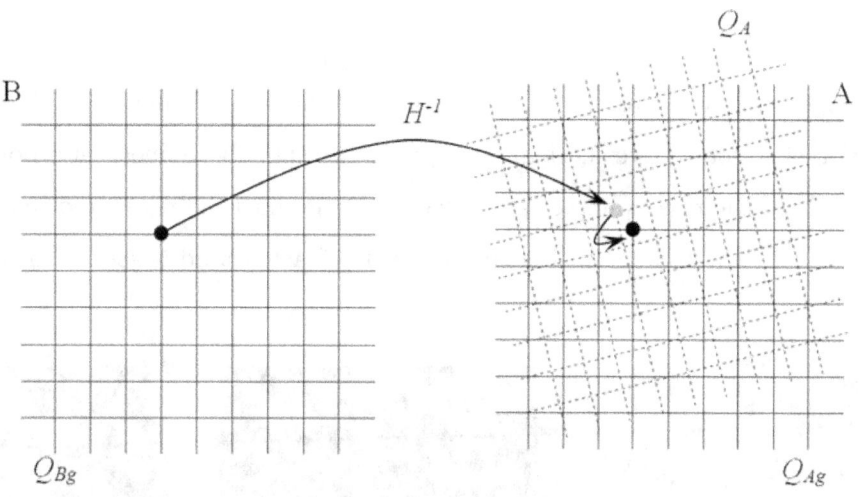

Figure 3.5: Integer and non integer coordinates systems

Thus, in order to perform the registration of our image, a black-and-white template reference image (B) is used to identify the homogeneous integer coordinates system Q_{Bg}. Specifically, considering a set P_B of 11 base points on the reference image B, a corresponding set P_A of 11 input points was located on each hand (sensed images A). The chosen landmarks location is depicted in Figure 3.6; it allows to identify the fingers and the centroid of the hand.

In this case, it is possible to estimate the mapping function H and compute its inverse H^{-1} to project the homogeneous integer coordinates Q_{Bg} (on the reference image B) onto

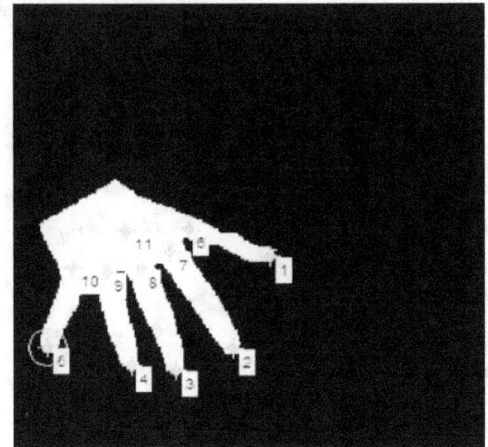 

Figure 3.6: Sensed image A (left) and reference image B (right) with their control points

the homogeneous non-integer coordinates Q_A on the sensed image A. Finally, the homogeneous integer coordinates Q_{Ag} are computed through the bilinear interpolation method. An example of the resampling of a segmented image into the registered image is shown in the Figure 3.7, where each pixel of the segmented background is set to NaN or 0.

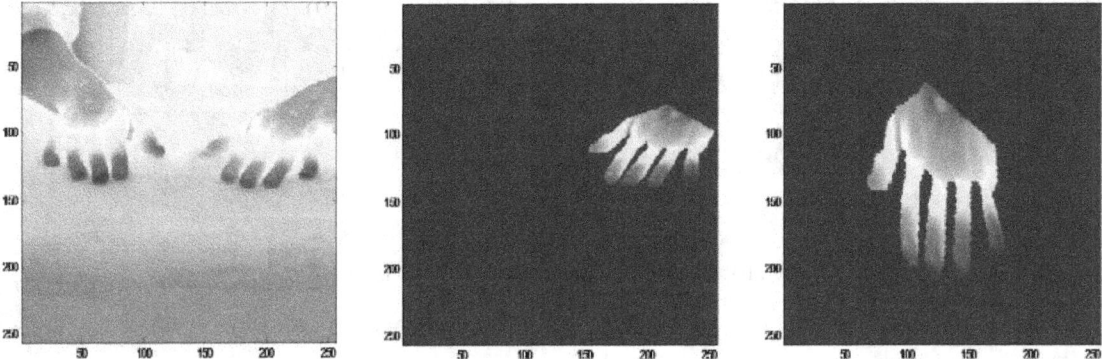

Figure 3.7: A typical example of IR image: original image (left), segmented left-hand image (middle), registered image (right)

Furthermore, notice that in order to make the images spatially homogeneous, a reflection of the left hand is needed (so to place it in the same position as the right one) respect to its own longitudinal central axis[3] (Figure 3.8).

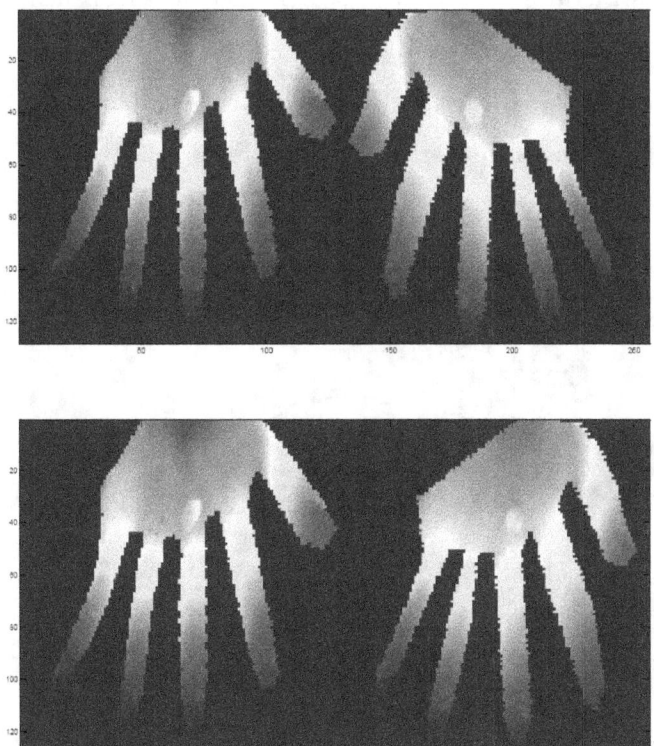

Figure 3.8: Example of image reflection: registered image (top row), reflected image (bottom row)

3.2 Feature Extraction

With the aim of developing automatic discrimination techniques for HCs, PRP and SSc people, we have to extract a set of features from the registered images. Such images display complex patterns at various orientations and we thus expect quite distinct texture characteristics among the classes. Texture analysis can be done either studying the point properties

---

[3] A reflection is a special case of magnification so that the transformation matrix is

$$H = \begin{bmatrix} -1 & 0 & 0 \\ 0 & 1 & 0 \\ 0 & 0 & 1 \end{bmatrix}.$$

of an image, in a pixel-based view, or explicitly defining the primitives that characterise the image, in a structural approach, to search their features such as spatial arrangement. In this section, we describe in detail two procedures. In the first one, since the images are not stationary in time and in space (Figure 3.9), we assume that the temperature values are realizations of a Gaussian Markov Random Field (GMRF) with time varying, mean $\boldsymbol{\mu}$ and covariance matrix $\mathbf{A}(\beta)$.

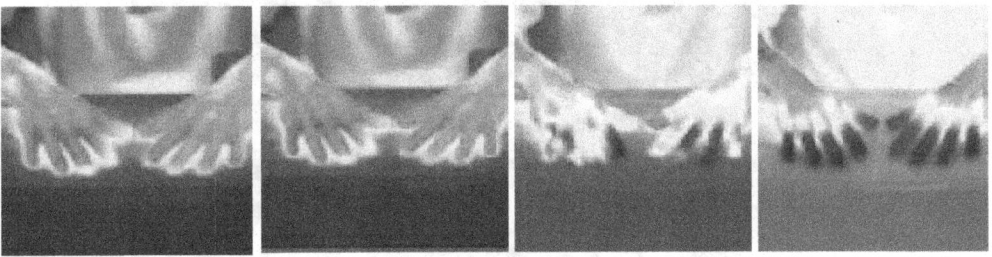

Figure 3.9: Examples of IR images

In the second procedure, other features of interest are obtained by extracting the information from co-occurrence matrices (CMs).

3.2.1 Gaussian Markov Random Fields

Consider a temporal sequence of $N \times M$ images and let $X(\mathbf{p})$, where $\mathbf{p} = (\mathbf{s}, t)$, a random variable representing the temperature value in Celsius degrees at time t at pixel of coordinates $\mathbf{s} = (s_x, s_y)$ with $s_x = 1, 2, \ldots, M$, $s_y = 1, 2, \ldots, N$, $t = 1, 2, \ldots, T$.

We write $\mathbf{X} = (X(\mathbf{p}_1), X(\mathbf{p}_2), \ldots, X(\mathbf{p}_g))'$ and take $g = N \times M \times T$. We will assume that the temperature values of the hands are considered as a realization of a Gaussian process

$$\mathbf{X} = \{X(\mathbf{p}), \mathbf{p} \in \Delta \subset \mathbb{R}^d\}, \ d = d_s + d_t, \qquad (3.2.1)$$

characterised by a $(g \times 1)$ mean vector, $\boldsymbol{\mu}$, and a $(g \times g)$ covariance matrix, $\boldsymbol{\Sigma}$ and the residual process $\varepsilon(\mathbf{p}) = X(\mathbf{p}) - \mu(\mathbf{p})$ is a zero mean stationary Gaussian process, with covariance function given by $\sigma(\mathbf{h}, k) = Cov[X(\mathbf{s}, t), X(\mathbf{s} + \mathbf{h}, t + k)]$, where $\mathbf{h} = \mathbf{s}_i - \mathbf{s}_j$ and $k = t_i - t_j$. If $\sigma(\mathbf{h}, k) = \sigma(||\mathbf{h}||, k)$, where $||\mathbf{h}|| = \sqrt{\mathbf{h}'\mathbf{h}}$, the process 3.2.1 is said to be *isotropic*.

In particular for $d_s = 2$ and $d_t = 1$, $\{X(\mathbf{s},t), \mathbf{s} \in \Delta_S \subset \mathbb{R}^2, t \in \Delta_T \subset \mathbb{R}^+\}$ is a stochastic spatio-temporal process, while for $d_s = 2$ and $d_t = 0$, $\{X(\mathbf{s}), \mathbf{s} \in \Delta_S \subset \mathbb{R}^2\}$ is a stochastic spatial process.

The mean structure can be modelled through a linear combination of independent variables with unknown parameters $\mathbf{b}$; i.e. $\boldsymbol{\mu} = \mathbf{Db}$ where the entries of the design matrix $\mathbf{D}$ are expressed as a function of the coordinates of site $\mathbf{p}$. Dealing with huge data sets, as in our case, it may be better for computational purposes to assume a conditional specification of the process. Therefore, the temperature levels are distributed according to a homogeneous Gaussian Markov Random Field (GMRF) if the distribution of X is multivariate normal with conditional means and conditional variances

$$E\left(X(\mathbf{p}_i)|X(\mathbf{p}_j), j \neq i\right) = \mu(\mathbf{p}_i) + \sum_{j \neq i} \beta_{ij} \left[X(\mathbf{p}_j) - \mu(\mathbf{p}_j)\right]$$

$$Var\left(X(\mathbf{p}_i)|X(\mathbf{p}_j), j \neq i\right) = v^2$$

where β_{ij} are the interaction parameters.

For a homogeneous process, and for a displacement vector $\mathbf{r}$, the conditional mean can also be rewritten as

$$E\left(X(\mathbf{p}_i)|X(\mathbf{p}_j), j \neq i\right) = \mu(\mathbf{p}_i) + \sum_{\mathbf{r} \in \delta_i^\alpha} \beta_{\mathbf{r}} \left[X(\mathbf{p}_i + \mathbf{r}) - \mu(\mathbf{p}_i + \mathbf{r})\right]$$

where δ_i^α is the set of neighbours of pixel $\mathbf{p}_i$, α is the GMRF *order* (cf 3.2.1.1) which is defined on a rectangular lattice by a given maximum distance between two pixels, $\beta_{\mathbf{r}} = \beta_{-\mathbf{r}}$, $\beta_{\mathbf{0}} = 0$ and $\beta_{\mathbf{r}} = 0, \forall \mathbf{r} : \mathbf{p}_i + \mathbf{r} \notin \delta_i^\alpha$.

This description is known as the conditional specification of a homogeneous GMRF [Dryden et al., 2002] and we say that $\mathbf{X} \sim N(\mathbf{Db}, \boldsymbol{\Sigma})$, with $\boldsymbol{\Sigma} = v^2 \mathbf{A}(\boldsymbol{\beta})^{-1}$, where $\mathbf{A}(\boldsymbol{\beta})$ is the $g \times g$ *potential matrix*, with entries equal to 1 along the main diagonal, the inverse correlations $-\beta_{ij}$ if the sites $\mathbf{p}_i$ and $\mathbf{p}_j$ are neighbours, and zero otherwise. Note that $\mathbf{A}(\boldsymbol{\beta})$ must be strictly positive definite for a non-degenerate distribution.

The first step is to estimate the parameters of the vector $\boldsymbol{\beta}$ containing the distinct $\beta_{\mathbf{r}}$ parameters, v^2 (the conditional variance) and the vector of trend parameters, $\mathbf{b}$. Given a realization $\mathbf{x}$ of the GMRF $\mathbf{X} \sim N\left(\mathbf{Db}, v^2 \mathbf{A}(\boldsymbol{\beta})^{-1}\right)$, the estimation of the parameter vector $\boldsymbol{\eta} = (\mathbf{b}', \boldsymbol{\beta}', v^2)$ can be obtained by the minimization of the negative log-likelihood

$$L(\boldsymbol{\eta}) = -\frac{g}{2}log(2\pi v^2) + \frac{1}{2}log(|\mathbf{A}(\boldsymbol{\beta})|) - \frac{1}{2v^2}(\mathbf{x} - \mathbf{Db})'\mathbf{A}(\boldsymbol{\beta})(\mathbf{x} - \mathbf{Db}). \tag{3.2.2}$$

The negative log-likelihood can be minimised in stages [Cressie, 1993]:

- For a fixed $\mathbf{A}(\boldsymbol{\beta})$

$$\hat{\mathbf{b}} = (\mathbf{D}'\mathbf{A}(\boldsymbol{\beta})\mathbf{D})^{-1}\mathbf{D}'\mathbf{A}(\boldsymbol{\beta})\mathbf{x} \tag{3.2.3}$$

$$\hat{v}^2 = \frac{1}{g}(\mathbf{x} - \mathbf{Db})'\mathbf{A}(\boldsymbol{\beta})(\mathbf{x} - \mathbf{Db}) \tag{3.2.4}$$

are the m.l. estimators of the trend parameters $\mathbf{b}$ and the conditional variance v^2.

- substituting (3.2.3) and (3.2.4) back into equation (3.2.2), the m.l. estimators of the spatial interaction parameters can be obtained by minimizing the negative log *profile likelihood*

$$L^*(\boldsymbol{\beta}) = \frac{g}{2}(log(2\pi) + 1) - \frac{1}{2}log(|\mathbf{A}(\boldsymbol{\beta})|) +$$
$$+ \frac{g}{2}log\left[\mathbf{x}'\mathbf{A}(\boldsymbol{\beta})\left\{\mathbf{I} - \mathbf{D}(\mathbf{D}'\mathbf{A}(\boldsymbol{\beta})\mathbf{D})^{-1}\mathbf{D}'(\mathbf{A}(\boldsymbol{\beta}))\right\}\mathbf{x}\right] \tag{3.2.5}$$

with respect to the unknown spatial interaction parameters β.

3.2.1.1 Neighbourhood Structure, Boundary Conditions and Toroidal MLE Estimation

In the conditional specification of a GMRF, it is fundamental to define a neighbourhood structure for the model. In particular, let a site B of integer coordinates (s_x, s_y) ($s_x \in \{1, 2, \ldots, M\}$, $s_y \in \{1, 2, \ldots, N\}$) on a $N \times M$ lattice (see 3.1).

$(s_x - 1, s_y + 1)$	$(s_x, s_y + 1)$	$(s_x + 1, s_y + 1)$
$(s_x - 1, s_y)$	$B \equiv (s_x, s_y)$	$(s_x + 1, s_y)$
$(s_x - 1, s_y - 1)$	$(s_x, s_y - 1)$	$(s_x + 1, s_y - 1)$

Table 3.1: Example of spatial lattice

Thus, we can define neighbourhood structures of several orders. In the table 3.2 are represented neighbours of the site B up to order 5.

o	o	o	o	o	o	o
o	5	4	3	4	5	o
o	4	2	1	2	4	o
o	3	1	B	1	3	o
o	4	2	1	2	4	o
o	5	4	3	4	5	o
o	o	o	o	o	o	o

Table 3.2: Neighbours of the site B up to order 5

Corresponding to the neighbourhood structure order, we have a vector $\mathbf{r} = (d_x, d_y)$ which defines the displacement from the site B. For example, a first order neighbourhood structure is defined by 4 displacements: $\mathbf{r} = (0, 1)$ and $\mathbf{r} = (0, -1)$ for neighbours which are one pixel apart vertically and $\mathbf{r} = (1, 0)$ and $\mathbf{r} = (-1, 0)$ for neighbours which are one pixel apart horizontally. While, a second order neighbourhood structure is defined by $\mathbf{r} = (0, 1)$, $\mathbf{r} = (0, -1)$, $\mathbf{r} = (1, 0)$ and $\mathbf{r} = (-1, 0)$ together with $\mathbf{r} = (1, 1)$ and $\mathbf{r} = (-1, -1)$ for diagonally adjacent neighbours in the north-east and south-west directions, resepectively and $\mathbf{r} = (-1, 1)$ and $\mathbf{r} = (1, -1)$ for diagonally adjacent neighbours in the north-west and south-east directions, respectively.

The neighbourhood structure of a spatio-temporal GMRF can be defined similarly, with times t considered as an extra dimension. The table 3.3 represents the displacements $\mathbf{r} = (d_x, d_y, d_t)$ from a site B of coordinates (s_x, s_y, t) $(t = 1, 2, \ldots, T)$ in a 3-dimensional lattice with a neighbourhood structure up to order 2.

t			$t+1$		
$(-1, 1, 0)$	$(0, 1, 0)$	$(1, 1, 0)$	$(-1, 1, 1)$	$(0, 1, 1)$	$(1, 1, 1)$
$(-1, 0, 0)$	B	$(1, 0, 0)$	$(-1, 0, 1)$	$(0, 0, 1)$	$(1, 0, 1)$
$(-1, -1, 0)$	$(0, -1, 0)$	$(1, -1, 0)$	$(-1, -1, 1)$	$(0, -1, 1)$	$(1, -1, 1)$

Table 3.3: Displacements $\mathbf{r}$ from a site B on a spatio-temporal lattice

Thus, for a first order neighbourhood structure we have 2 displacement vectors $\mathbf{r} = (1, 0, 0)$, $\mathbf{r} = (-1, 0, 0)$ for neighbours which are one pixel apart horizontally at a given time t; 2 displacement vectors $\mathbf{r} = (0, 1, 0)$, $\mathbf{r} = (0, -1, 0)$ for neighbours which are one pixel apart vertically at a given time t; 2 displacement vectors $\mathbf{r} = (0, 0, 1)$, $\mathbf{r} = (0, 0, -1)$ for neighbours which are at the same pixel but at different times, $t - 1$ or $t + 1$. Similarly, for the second order neighbours, we can identify, together with the first order displacement vectors, diagonally displacement vectors at a given time t.

An alternative strategy to define the order of a GMRF is to specify the space-time neighbourhood from a time series perspective and in this case the order is a vector $\alpha = (\alpha_s, \alpha_t)$ consisting of the spatial and the temporal lags, respectively [Fontanella et al., 2008]. In general, we will have $(1 + 2\alpha_1)(1 + \alpha_2)$ parameters for a homogeneous process, and $(1 + \alpha_1)(1 + \alpha_2)$ parameters for a completely symmetric one (isotropic).

An important part of the GMRF model specification is the choice of boundary conditions (b.c.) for a stationary process, since elements of $\boldsymbol{\Sigma}^{-1}$ for boundary sites on a finite lattice can be very complicated [Besag and Moran, 1975]. In general, to deal with GMRFs on finite rectangular lattices, the most convenient b.c. are toroidal b.c. These specify that each dimension is assumed to be wrapped around, so that the first and last coordinates are adjacent (see Figure 3.10).

If toroidal b.c. are assumed, the estimation of the parameter vector $\boldsymbol{\eta} = (\mathbf{b}', \boldsymbol{\beta}', v^2)$ for the process 3.2.1 is easier because $\mathbf{A}(\boldsymbol{\beta})$ is block circulant of order d (e.g., if $d = 3$, $\mathbf{A}(\boldsymbol{\beta})$ is block circulant and each block is itself block circulant).

For example, consider a homogeneous GMRF of order $\boldsymbol{\alpha} = (\alpha_s, \alpha_t) = (2, 1)$ and assume toroidal b.c. The matrix $\mathbf{A}(\boldsymbol{\beta})$

$$\mathbf{A}(\boldsymbol{\beta}) = \begin{pmatrix} \mathbf{A}_1 & \mathbf{A}_2 & \mathbf{0} & \cdots & \cdots & \mathbf{0} & \cdots & \mathbf{0} & \mathbf{0} & \mathbf{A}_T \\ \mathbf{A}_T & \mathbf{A}_1 & \mathbf{A}_2 & \mathbf{0} & \cdots & \mathbf{0} & \cdots & \mathbf{0} & \mathbf{0} & \mathbf{0} \\ \mathbf{0} & \mathbf{A}_T & \mathbf{A}_1 & \mathbf{A}_2 & \cdots & \mathbf{0} & \cdots & \mathbf{0} & \mathbf{0} & \mathbf{0} \\ \vdots & \vdots & \vdots & \cdots & \cdots & \vdots & \cdots & \vdots & \vdots & \vdots \\ \mathbf{0} & \mathbf{0} & \mathbf{0} & \cdots & \cdots & \mathbf{0} & \cdots & \mathbf{A}_T & \mathbf{A}_1 & \mathbf{A}_2 \\ \mathbf{A}_2 & \mathbf{0} & \mathbf{0} & \cdots & \cdots & \mathbf{0} & \cdots & \mathbf{0} & \mathbf{A}_T & \mathbf{A}_1 \end{pmatrix}$$

is a $g \times g$ ($g = N \times M \times T$) block-circulant matrix of level three and it can be divided into T different blocks of order $(N \times M) \times (N \times M)$ with block-circulant structure. Each of the T blocks can be subdivided into N different blocks, each of which is a circulant of order M. In particular $\mathbf{A}_1$ contains the spatial parameters $\beta_{(0,1,0)}$, $\beta_{(1,0,0)}$, $\beta_{(1,-1,0)}$ and $\beta_{(-1,1,0)}$

X(N-1,M-2)	X(N-1,M-1)	X(N-1,1)	X(N-1,2)	......	X(N-1,M-2)	X(N-1,M-1)	X(N-1,1)	X(N-1,2)
X(n,M-2)	X(N,M)	X(N,1)	X(N,2)	......	X(n,M-2)	X(N,M)	X(N,1)	X(N,2)
X(1,M-1)	X(1,M)	**X(1,1)**	**X(1,2)**	......	**X(1,M-1)**	**X(1,M)**	X(1,1)	X(1,2)
X(2,M-2)	X(2,M)	**X(2,1)**	**X(2,2)**	......	**X(2,M-2)**	**X(2,M)**	X(2,1)	X(2,2)
......	......	......	......	......	......	......	......	......
X(N-1,M-2)	X(N-1,M-1)	**X(N-1,1)**	**X(N-1,2)**	......	**X(N-1,M-2)**	**X(N-1,M-1)**	X(N-1,1)	X(N-1,2)
X(n,M-2)	X(N,M)	**X(N,1)**	**X(N,2)**	......	**X(n,M-2)**	**X(N,M)**	X(N,1)	X(N,2)
X(1,M-1)	X(1,M)	X(1,1)	X(1,2)	......	X(1,M-1)	X(1,M)	X(1,1)	X(1,2)
X(2,M-2)	X(2,M)	X(2,1)	X(2,2)	......	X(2,M-2)	X(2,M)	X(2,1)	X(2,2)

Figure 3.10: Example of toroidal b.c. on a $N \times M$ lattice

in the following circulant matrices

$$\mathbf{A}_{1,1} = circ(1, -\beta_{(1,0,0)}, 0, \cdots, 0, -\beta_{(1,0,0)})$$
$$\mathbf{A}_{1,2} = circ(-\beta_{(0,-1,0)}, -\beta_{(1,-1,0)}, 0, \cdots, 0, -\beta_{(-1,1,0)})$$
$$\mathbf{A}_{1,N} = circ(-\beta_{(0,1,0)}, \beta_{(-1,1,0)}, 0, \cdots, 0, \beta_{(1,-1,0)}).$$

The matrices $\mathbf{A}_2$ and $\mathbf{A}_T$ contain the temporal parameter $\beta_{(0,0,1)}$ and the spatio-temporal parameters $\beta_{(0,1,1)}$, $\beta_{(1,0,1)}$, $\beta_{(1,-1,1)}$ and $\beta_{(-1,1,1)}$ in the following circulant matrices

$$\mathbf{A}_{2,1} = circ(-\beta_{(0,0,1)}, -\beta_{(1,0,1)}, 0, \cdots, 0, -\beta_{(1,0,1)})$$
$$\mathbf{A}_{2,2} = circ(-\beta_{(0,-1,1)}, -\beta_{(1,-1,1)}, 0, \cdots, 0, -\beta_{(-1,1,1)})$$
$$\mathbf{A}_{2,N} = circ(-\beta_{(0,1,1)}, \beta_{(-1,1,1)}, 0, \cdots, 0, \beta_{(1,-1,1)}).$$

Block circulant matrices with circulant blocks of any level (as well as circulant matrices themselves) have the important property that all the information is contained in the first row (column). Thus, under toroidal b.c. assumption, $\mathbf{A}(\boldsymbol{\beta})$ can be easily diagonalised with a d-dimensional fast Fourier transform [Besag and Moran, 1975]. For example, in the bi-dimensional spatial framework where $\mathbf{p} = (s_x, s_y)$ and $d_s = 2$, $d_t = 0$, if $\mathbf{P}_j$ is the usual discrete Fourier transform matrix, then $\mathbf{P} = \mathbf{P}_N \otimes \mathbf{P}_M$ is the $n \times n$ matrix of the eigenvectors of $\mathbf{A}(\boldsymbol{\beta})$, with $\otimes$ being the Kronecker product. The eigenvalues of $\mathbf{A}(\boldsymbol{\beta})$ are simple to obtain from the 2D discrete Fourier transform and we write the eigenvalues as $q_1, \ldots, q_n$. Hence, $\mathbf{A}(\boldsymbol{\beta}) = \mathbf{P}\mathbf{Q}\mathbf{P}^T$, where $\mathbf{Q} = diag(q_1, \ldots, q_n)^T$.

Similarly, in the spatio-temporal framework where $\mathbf{p} = (\mathbf{s}, t)$ and $d_s = 2$, $d_t = 1$, if $\mathbf{P}_j$ is the usual discrete Fourier transform matrix, then $\mathbf{P} = \mathbf{P}_N \otimes \mathbf{P}_M \otimes \mathbf{P}_T$ is the $g \times g$ matrix of the eigenvectors of $\mathbf{A}(\boldsymbol{\beta})$. The eigenvalues of $\mathbf{A}(\boldsymbol{\beta})$ result from the 3D discrete Fourier transform and we write the eigenvalues as $q_1, \ldots, q_n$. Hence, $\mathbf{A}(\boldsymbol{\beta}) = \mathbf{P}\mathbf{Q}\mathbf{P}^T$, where $\mathbf{Q} = diag(q_1, \ldots, q_n)^T$. The resulting log-likelihood is

$$L(\mathbf{b}, v^2) = -\frac{g}{2}log(2\pi\, v^2) + \frac{1}{2}\sum_{i=1}^{g}\log(q_i) - \frac{1}{2v^2}(\mathbf{x} - \mathbf{D}\mathbf{b})^T \mathbf{P}\mathbf{Q}\mathbf{P}^T(\mathbf{x} - \mathbf{D}\mathbf{b})$$

and the maximization can be carried out with only $O(g \log g)$ steps [Besag and Moran, 1975, Dryden et al., 2002] over the valid parameter space, i.e. subject to $\mathbf{A}(\boldsymbol{\beta})$ being strictly positive definite.

3.2.2 Texture Analysis

Image analysis techniques have played an important role in several medical applications. In general, we are interested in automatically extracting features from the image which capture morphological, color or certain textural properties, used for classification tasks. The textural properties computed are closely related to the application domain to be used. For example, Sutton and Hall [Sutton and Hall, 1972] discuss the classification of pulmonary disease using texture features and looking to their changes in X-ray images as opposed to clearly delineated lesions; Landeweerd and Gelsema [Landeweerd and Gelsema, 1978] extracted various first order (e.g. means computed on certain grey levels regions of interest) and second order statistics (e.g. gray levels co-occurrence matrices) to differentiate different types of white blood cells.

Here, we use a pixel-based approach to identify further basic patterns that could represent the natural texture structure of the RP. Specifically, we perform a texture analysis by extracting information in the form of a co-occurrence matrix (CM) and by summarizing this information through the calculation of some measures of texture on the CM [Cocquerez and Philipp, 1995]. To calculate these measures, for each image we first classify the estimated stationary residual process, $\hat{\varepsilon}_{ij} = x_{ij}\hat{\mu}_{ij}$, in L levels, considering the quantiles of order L of the $\hat{\varepsilon}$ distribution. Thus, we obtain the classified image $\{I_L(x, y) \in \{1, 2, \ldots, L\} : x \in \{1, 2, \ldots, M\}, y \in \{1, 2, \ldots, N\}\}$. Then, for a given spatial displacement vector, $\mathbf{r}$, which defines pairs of neighbours in the spatial domain, we compute the CM which provides a tabulation of how often different combinations of classified pixel values occur in an image [Cocquerez and Philipp, 1995]. The Figure 3.11 shows an example of co-occurrence counting for a 8-levels classified image (left) and the corresponding co-occurrence matrix (right) for the displacement vector $\mathbf{r} = (1, 0)$ (i.e. a pixel is a neighbour of the one to its right side).

More specifically, the (i, j)-th element of the $(L \times L)$ CM, denoted here as $\tilde{\mathbf{C}}_{\mathbf{r}}$, represents the frequency, $n(i, j)$, of occurrence of a pair of classified pixel values, separated by the displacement $\mathbf{r}$ and having temperature levels i and j, respectively. It is also useful to normalise the $\tilde{\mathbf{C}}_{\mathbf{r}}$ by relative frequencies[4], $f(i, j)$.

---

[4]In this case, the $f(i, j)$ also represents, at two arbitrary locations separated by the displacement $\mathbf{r}$, an estimate of the probability of observing temperature levels i and j. Under this assumption, we are in a pure

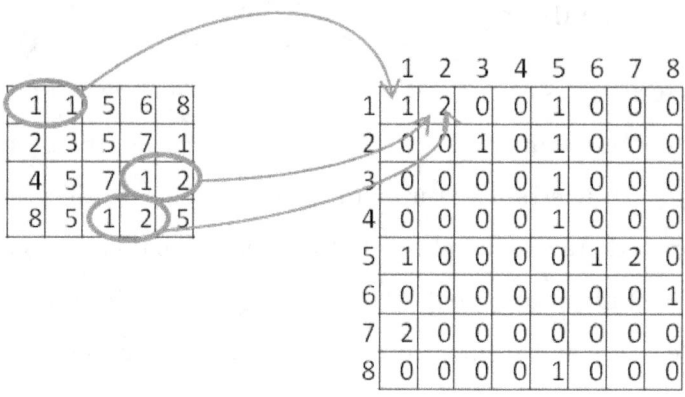

Figure 3.11: Examples of co-occurrence counting

Formally, the number $n(i,j)$ of occurrences of pairs of levels i and j which are a distance $\mathbf{r} = (d_x, d_y)$ apart, is given as

$$\tilde{\mathbf{C}}_{\mathbf{r}}(i,j) = |\{((m,s),(t,v)) : I_L(r,s) = i, I_L(t,v) = j\}|$$

where $(m,s), (t,v) \in N \times M$, $(t,v) = (m + d_x, s + d_y)$, I_L denote an $N \times M$ image with L levels and $|\cdot|$ is the cardinality of a set. Therefore, letting

$$f(i,j) = \frac{\tilde{\mathbf{C}}_{\mathbf{r}}(i,j)}{\sum_i \sum_j \tilde{\mathbf{C}}_{\mathbf{r}}(i,j)},$$

we have the normalised CM, $\mathbf{C_r}$ (see Table 3.4).

Our initial assumption in characterizing image texture, is that all the texture information is contained in the CM. Hence, all the textural features we suggest are extracted from the CM. Specifically, Haralick *et alii* ([Haralick et al., 1973]) propose a set of textural information meausers from wich we compute, for a displacement vector $\mathbf{r}$, the following:

$$T_1(\mathbf{r}) = \sum_{i,j} (i-j)^2 f(i,j) \qquad (3.2.6)$$

$$T_2(\mathbf{r}) = \sum_{i,j} \frac{f(i,j)}{1 + |i-j|} \qquad (3.2.7)$$

---

statistical information theory framework.

	1	2	...	...	j	...	...	L	
1					$f(1,j)$				$f(1,\cdot)$
2					$f(2,j)$				$f(2,\cdot)$
⋮					⋮				⋮
⋮					⋮				⋮
i	$f(i,1)$	...	...	...	$f(i,j)$	...	...	$f(i,L)$	$f(i,\cdot)$
⋮					⋮				⋮
⋮					⋮				⋮
L					$f(L,j)$				$f(L,\cdot)$
	$f(\cdot,1)$	$f(\cdot,2)$	...	...	$f(\cdot,j)$	...	...	$f(\cdot,L)$	1

Table 3.4: Co-occurrence matrix $\mathbf{C_r}$

$$T_3(\mathbf{r}) = \sum_{i,j} f(i,j)^2 \tag{3.2.8}$$

$$T_4(\mathbf{r}) = \frac{\sum_{i,j} ij f(i,j) - \sum_i i f(i,\cdot) \sum_j j f(\cdot,j)}{\sigma_i\, \sigma_i} \tag{3.2.9}$$

$$T_5(\mathbf{r}) = \sum_{i,j} f(i,j) \log_2 \frac{f(i,j)}{f(i,\cdot)f(\cdot,j)} \tag{3.2.10}$$

where $\sigma_i = \left[\sum_i i^2 f(i,\cdot) - (\sum_i i f(i,\cdot))^2\right]^{1/2}$, $\sigma_j = \left[\sum_j j^2 f(\cdot,j) - (\sum_j j f(\cdot,j))^2\right]^{1/2}$ and $f(\cdot,j)$ and $f(i,\cdot)$ represent the marginal relative frequencies over the indices j and i, respectively. The indices T_1 and T_2 represent *Contrast* and *Homogeneity* measures and use weights related to the distance from the diagonal of the CM; T_3 is known as *Energy* and gives information about orderliness, meaning that a small value of T_3 corresponds to a CM with very few dominant (large) temperature levels. Finally, T_4 and T_5 are *Correlation* and *Mutual Information* indices, respectively; they provide a measure of the linear and non linear dependence of pairs of classified pixel values (Cf [Cover and Thomas, 1991, Haussler and Opper, 1997]).

We remark that, even though these features contain information about the textural characteristics of the image, it is hard to identify wich specific textural characteristics is

3.3 The Dataset

In this section we describe the data-matrix to be used in the classification task, both in spatial and spatio-temporal framework.

3.3.1 The Dataset in the Spatial Domain

Consider the k-th individual ($k = 1, 2, \ldots, 44$) and let $\mathbf{x}_h(k)$ the 256×256 processed image of temperature values of his hand h (where $h = 1, 2$ represent right and left hand, respectively) collected in the experiment at the time $t = 1$. As in section 3.2.1 with $N = M = 256$, let $\hat{\varepsilon}_h(k)$ the 256×256 image of the estimated residual for the k-th individual on each hand. In particular, for computational reasons, the processed images were cropped to dimension 128×128, simply defining a 128×128 region and removing the outer part (Cf [Gonzalez et al., 2004]).

A 6-parameter quadratic surface model was fitted to represent the trend temperature. The choice of the quadratic trend is supported by the visual exploratory analysis of the surface (Figure 3.12) and by comparing the R^2 values with those of a linear and cubic trends (Figure 3.13).

Thus, the residual correlation was modelled by a second order homogeneous GMRF, $\mathbf{X} \sim N\left(\mathbf{Db}, v^2 \mathbf{A}(\boldsymbol{\beta})^{-1}\right)$ where the $n \times 6$ design matrix $\mathbf{D}_h(k)$ of lattice coordinates system is as follow:

$$\mathbf{D}_h(k) = \begin{bmatrix} 1 & \cdots & \cdots & \cdots & \cdots & \cdots \\ \vdots & \vdots & \vdots & \vdots & \vdots & \vdots \\ 1 & s_x & s_y & s_x^2 & s_y^2 & s_x s_y \\ \vdots & \vdots & \vdots & \vdots & \vdots & \vdots \\ 1 & \cdots & \cdots & \cdots & \cdots & \cdots \end{bmatrix}. \tag{3.3.1}$$

For each individual and for each hand, we have a vector of features represented by the following parameters: $\hat{\boldsymbol{\eta}}_h(k) = (\hat{\mathbf{b}}'_h(k), \hat{\boldsymbol{\beta}}'_h(k), \hat{v}_h^2(k))$.

Specifically:

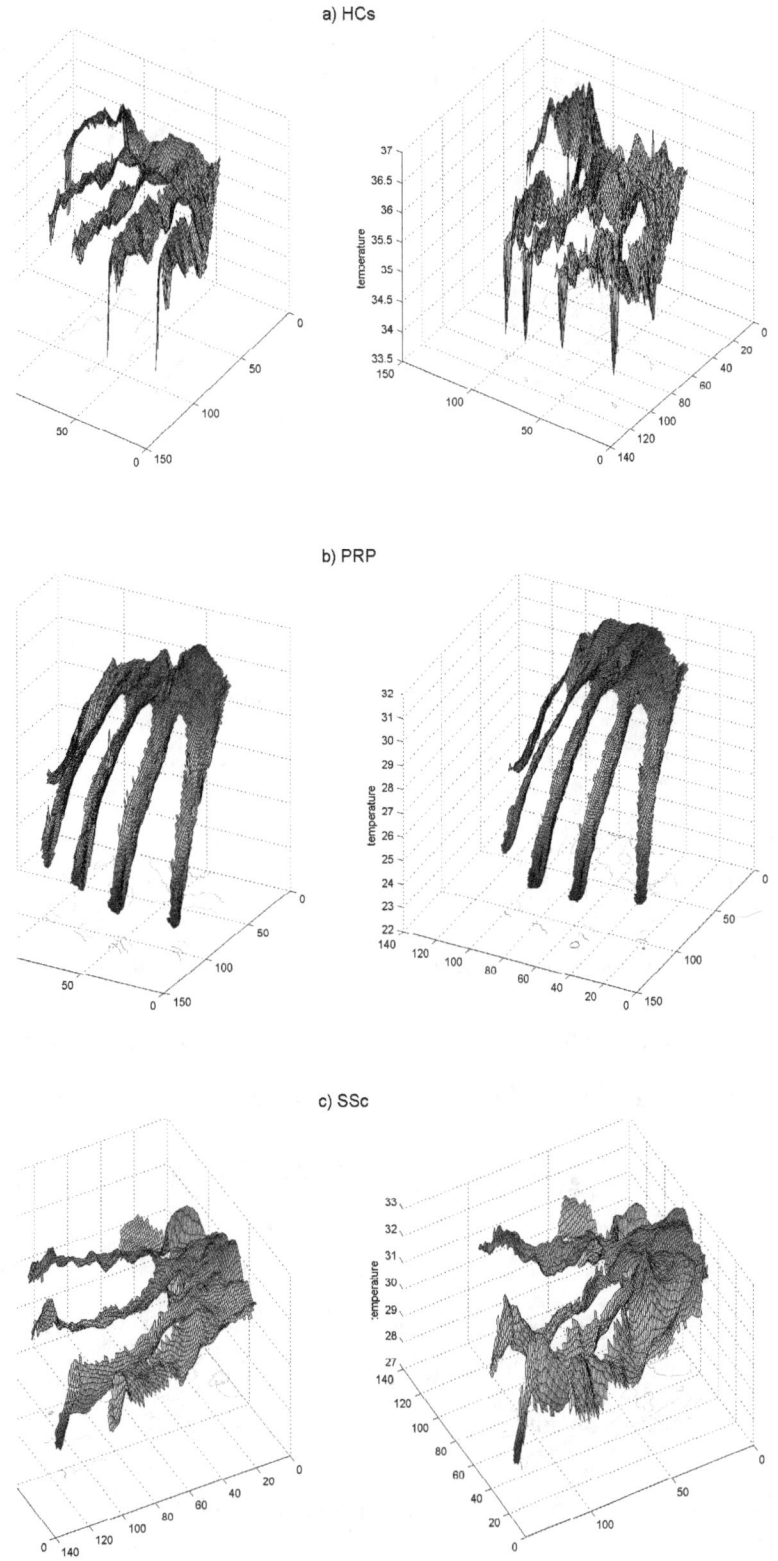

Figure 3.12: Surface of hand's temperature values for typical RP patients (HCs, PRP, SSc) on a 128×128 lattice

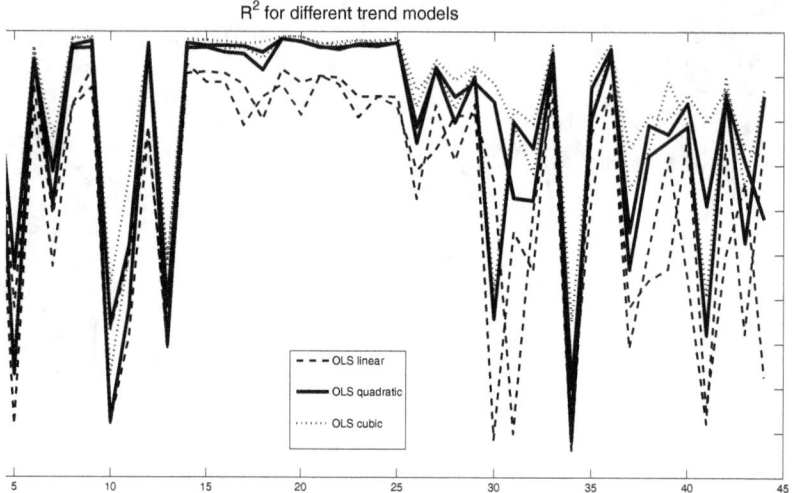

Figure 3.13: R^2 for different trend models

- $\hat{\mathbf{b}}_h(k) = (\hat{b}_{0h}(k), \hat{b}_{1h}(k), \hat{b}_{2h}(k), \hat{b}_{3h}(k), \hat{b}_{4h}(k), \hat{b}_{5h}(k))'$ is a 6×1 vector of the estimated trend parameters for the k-th individual relative to the $n \times 6$ design matrix $\mathbf{D}_h(k)$ of lattice coordinates system

- $\hat{\boldsymbol{\beta}}_h(k) = (\hat{\beta}_{(1,0)h}(k), \hat{\beta}_{(0,-1)h}(k), \hat{\beta}_{(1,-1)h}(k), \hat{\beta}_{(-1,-1)h}(k))'$ is the 4×1 vector of the spatial interaction parameters for the k-th individual.

- $\hat{v}_h^2(k)$ is the estimated conditional variance for the k-th individual.

Thus, as regards for GMRF, for the k-th individual, the vector of features, for both hands, is then represented by the following 1×22 vector of parameters:

$$\mathbf{g}(k) = (\hat{\mathbf{b}}_2'(k), \hat{\mathbf{b}}_1'(k), \hat{\boldsymbol{\beta}}_2'(k), \hat{\boldsymbol{\beta}}_1'(k), \hat{v}_2^2(k), \hat{v}_1^2(k)).$$

Let $\hat{\varepsilon}(k) = [\hat{\varepsilon}_1(k) \ \hat{\varepsilon}_2(k)]$, $k = 1, 2, \ldots, 44$ be the 256×256 matrix obtained by joining the two images of estimated residuals for both hands. According to the section 3.2.2, $\hat{\varepsilon}(k)$ is classified into the L levels image I_L. The number of levels, L, represent the quantiles of order L of the distribution

$$\hat{\varepsilon} = (vec(\hat{\varepsilon}_1(1))', vec(\hat{\varepsilon}_2(1))', vec(\hat{\varepsilon}_1(2))', vec(\hat{\varepsilon}_2(2))', \ldots, vec(\hat{\varepsilon}_1(44))', vec(\hat{\varepsilon}_2(44))')$$

and we tried $L = 10, 15, 20, 25$.

In particular, were considered $L = 10$ levels due to the results comparison of a linear discriminant analysis classification, using a stepwise feature selection procedure.

Under homogeneity assumptions, 32 different displacements, $\mathbf{r}_j$, $j = 1, 2, \ldots, 32$ (see Figure 3.14), were considered to define the neighbourhood structure.

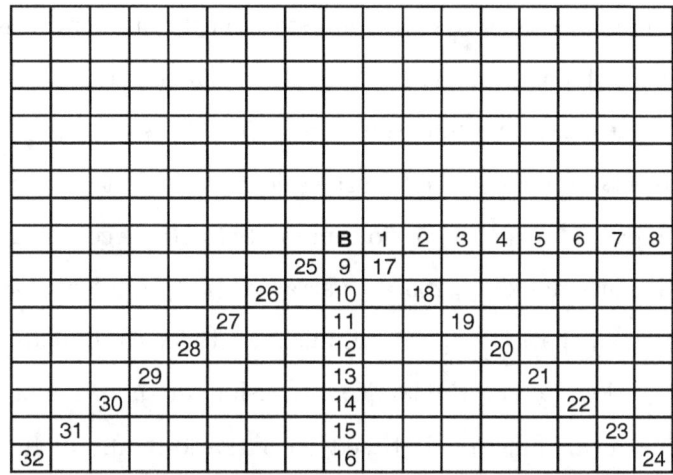

Figure 3.14: Displacements j for the co-occurrence matrices $\mathbf{C}_{\mathbf{r}_j}$

Therefore, for each individual, we have 32 symmetric co-occurrence matrices $\mathbf{C}_{\mathbf{r}_j}$ from which we compute different texture measures:

$$\mathbf{T}(\mathbf{r}_j, k) = (T_3(\mathbf{r}_j, k), T_5(\mathbf{r}_j, k), T_1(\mathbf{r}_j, k), T_4(\mathbf{r}_j, k), T_2(\mathbf{r}_j, k)).$$

This procedure generates the following 1×160 vector of features

$$\mathbf{t}(k) = (\mathbf{T}(\mathbf{r}_1, k), \mathbf{T}(\mathbf{r}_2, k), \ldots, \mathbf{T}(\mathbf{r}_{32}, k)).$$

By letting $\mathbf{c}(k) = (\mathbf{g}(k), \mathbf{t}(k))$, $k = 1, 2, \ldots, 44$, for classification pourposes we thus consider the 44×182 data matrix

$$\mathbf{A} = \begin{bmatrix} \mathbf{c}(1) \\ \mathbf{c}(2) \\ \vdots \\ \mathbf{c}(44) \end{bmatrix}.$$

3.3.2 The Dataset in the Spatio-Temporal Domain

As a natural extension of the previous section, consider the k-th individual ($k = 1, 2, \ldots, 44$) and, for a specific time point t, let $\mathbf{x}_h(k,t)$ the 256×256 processed image of temperature values for the left ($h = 1$) and right ($h = 2$) hands. Since the rewarming process varies slowly after the cold stress, we have downsampled the time series by considering one image per minute. Furthermore to by-pass the non stationarity problem caused by the cold-stress at $t = 5$, we focused the analysis on $t = 7, 9, 11, \ldots, 43$. For the k-th individual, we have a sequence of 19 images, each of dimension 256×256.

For each subject, as before, the identification of the feature variables starts with the estimation of the parameters of a spatio-temporal GMRF. According to section 3.2.1, consider the three-dimensional $256 \times 256 \times 19$ array, $\mathbf{x}_h(k,t)$ and let $\hat{\varepsilon}_h(k,t)$ the corresponding array of the estimated spatio-temporal residuals for the k-th individual for a specific hand. To reduce computational costs, the lattice dimension, at time t, is cropped to 128×128.

The estimated mean function is based on a spatio-temporal function expressed as a polynomial function of time and spatial coordinates; more specifically, we consider a trend which is linear in space and quadratic in time, including interaction terms between space and time. For the residual correlated process, we consider a spatio-temporal neighbourhood structure of order $\boldsymbol{\alpha} = (2,1)$ with 4 neighbours in space (homogeneous second order in space) and one lag in time. The spatio-temporal neighbourhood structure is shown in the Table 3.5 respect to a reference site B.

	t			$t+1$	
	B	$\beta_{(1,0,0)}$		$\beta_{(0,0,1)}$	$\beta_{(1,0,1)}$
$\beta_{(-1,-1,0)}$	$\beta_{(0,-1,0)}$	$\beta_{(1,-1,0)}$	$\beta_{(-1,-1,1)}$	$\beta_{(0,-1,1)}$	$\beta_{(1,-1,1)}$

Table 3.5: Spatio-temporal interaction parameters

Therefore, the corresponding GMRF is

$$\mathbf{X} \sim N\left(\mathbf{Db}, v^2 \mathbf{A}(\boldsymbol{\beta})^{-1}\right)$$

where $\mathbf{D}_h(k)$ represent the following $g \times 7$ design matrix of lattice space-time coordinates system

$$\mathbf{D}_h(k) = \begin{bmatrix} 1 & \cdots & \cdots & \cdots & \cdots & \cdots & \cdots \\ \vdots & \vdots & \vdots & \vdots & \vdots & \vdots & \vdots \\ 1 & t & t^2 & s_x & s_y & s_x t & s_y t \\ \vdots & \vdots & \vdots & \vdots & \vdots & \vdots & \vdots \\ 1 & \cdots & \cdots & \cdots & \cdots & \cdots & \cdots \end{bmatrix}.$$

The estimation $\hat{\boldsymbol{\eta}}_h(k) = (\hat{\mathbf{b}}'_h(k), \hat{\boldsymbol{\beta}}'_h(k), \hat{v}_h^2(k))$ is a 17 × 1 vector.

Specifically:

- $\hat{\mathbf{b}}_h(k) = (\hat{b}_{0h}(k), \hat{b}_{1h}(k), \hat{b}_{2h}(k), \hat{b}_{3h}(k), \hat{b}_{4h}(k), \hat{b}_{5h}(k), \hat{b}_{6h}(k))'$ is the 7 × 1 vector of the estimated trend parameters relative to the $g \times 6$ design matrix $\mathbf{D}_h(k)$;

- the interaction parameters vector of the GMRF, is

$$\hat{\boldsymbol{\beta}}_h(k) = \begin{bmatrix} \hat{\beta}_{(1,0,0)h}(k) \\ \hat{\beta}_{(0,-1,0)h}(k) \\ \hat{\beta}_{(1,-1,0)h}(k) \\ \hat{\beta}_{(-1,-1,0)h}(k) \\ \hat{\beta}_{(0,0,1)h}(k) \\ \hat{\beta}_{(1,0,1)h}(k) \\ \hat{\beta}_{(0,-1,1)h}(k) \\ \hat{\beta}_{(1,-1,1)h}(k) \\ \hat{\beta}_{(-1,-1,1)h}(k) \end{bmatrix}$$

where the vector components represent vertically, horizontally, diagonal South-East (or North-West) and diagonal North-East (or South-West) together with the time interaction parameters, rispectively;

- $\hat{v}_h^2(k)$ be the estimated conditional variance.

Thus, as regards for GMRF, for the k-th individual, the vector of features, for both hands, is then represented by the following 1 × 34 vector of parameters:

$$\mathbf{g}(k) = (\hat{\mathbf{b}}'_2(k), \hat{\mathbf{b}}'_1(k), \hat{\boldsymbol{\beta}}'_2(k), \hat{\boldsymbol{\beta}}'_1(k), \hat{v}_2^2(k), \hat{v}_1^2(k)).$$

Let $\hat{\varepsilon}(k) = [\hat{\varepsilon}_1(k)\ \hat{\varepsilon}_2(k)]$, $k = 1, 2, \ldots, 44$ be the $256 \times 256 \times 19$ array obtained by joining the two array of estimated residuals for both hands. According to the section 3.2.2 $\hat{\varepsilon}(k)$ is classified into the L levels image I_L. The number of levels, L, represent the quantiles of order L of the distribution

$$\hat{\varepsilon} = (vec(\hat{\varepsilon}_1(1))', vec(\hat{\varepsilon}_2(1))', vec(\hat{\varepsilon}_1(2))', vec(\hat{\varepsilon}_2(2))', \ldots, vec(\hat{\varepsilon}_1(44))', vec(\hat{\varepsilon}_2(44))')$$

In particular, as in the spatial framework, were considered $L = 10$. Under homogeneity assumptions, 49 different displacements, $\mathbf{r}_j$, $j = 1, 2, \ldots, 49$ (see Figure 3.15), were considered to define the neighbourhood structure. Therefore, for the k-th individual, we have 49

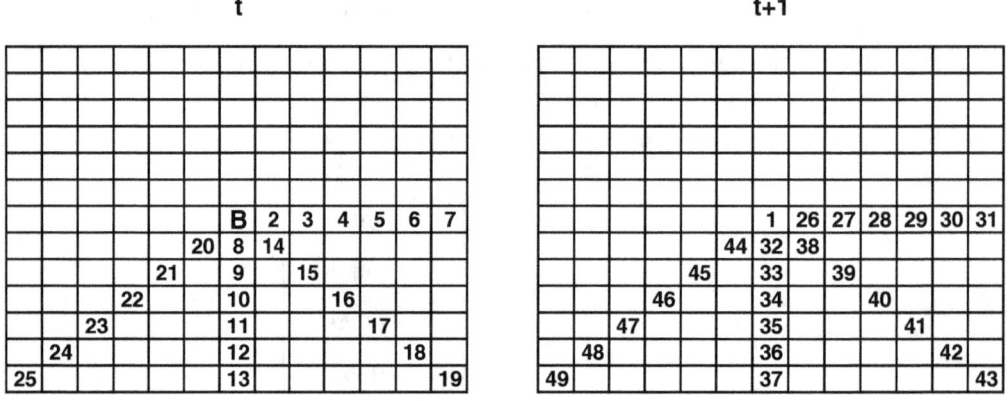

Figure 3.15: Spatio-temporal displacements j for the co-occurrence matrices $\mathbf{C}_{\mathbf{r}_j}$

symmetric co-occurence matrices $\mathbf{C}_{\mathbf{r}_j}$ from which we compute different texture measures:

$$\mathbf{T}(\mathbf{r}_j, k) = (T_3(\mathbf{r}_j, k), T_5(\mathbf{r}_j, k), T_1(\mathbf{r}_j, k), T_4(\mathbf{r}_j, k), T_2(\mathbf{r}_j, k)).$$

This procedure generates the following 1×245 vector of features

$$\mathbf{t}(k) = (\mathbf{T}(\mathbf{r}_1, k), \mathbf{T}(\mathbf{r}_2, k), \ldots, \mathbf{T}(\mathbf{r}_{49}, k)).$$

By letting $\mathbf{c}(k) = (\mathbf{g}(k), \mathbf{t}(k))$, $k = 1, 2, \ldots, 44$, for classification pourposes we thus consider the 44×279 data matrix

$$\mathbf{A} = \begin{bmatrix} \mathbf{c}(1) \\ \mathbf{c}(2) \\ \vdots \\ \mathbf{c}(44) \end{bmatrix}.$$

Chapter 4

Fisher's Linear Discriminant Analysis for the High Dimension/Small Sample Size Problems

Fisher's linear discrimant analysis (FLDA) is one of the prominent methods in supervised classification tasks. Even in the presence of today's necessity to handle high dimensional data, FLDA is still widely used by researchers. However, classification based on FLDA is challenging when the number of variables exceeds the number of given objects due to the singularity of the covariance matrices. Specifically, consider a classical FLDA task with G groups and $n = \sum_{j=1}^{G} n_j$ training objects $(\mathbf{x}_i, y_i)$ with $\mathbf{x}_i = (x_{i1}, x_{i2}, \ldots, x_{ip}) \in \mathbb{R}^p$, $y_i \in \{1, 2, \ldots, G\}$ and n_j representing the size of group j. Then, *Fisher's criterion* [Fisher, 1936] is to find, subsequently, at most $G - 1$ transformation vectors $\mathbf{a}$ that have maximal separation ratio by solving the maximization problem

$$max_{\mathbf{a} \in \mathbb{R}^p - \{0\}} \frac{\mathbf{a}^T \mathbf{B} \mathbf{a}}{\mathbf{a}^T \mathbf{W} \mathbf{a}} \tag{4.0.1}$$

where

$$\mathbf{B} = \frac{1}{G-1} \sum_{j=1}^{G} n_j (\bar{\mathbf{x}}_j - \bar{\mathbf{x}})(\bar{\mathbf{x}}_j - \bar{\mathbf{x}})^T = \frac{\mathbf{H}}{G-1} \tag{4.0.2}$$

$$\mathbf{W} = \frac{1}{n-G} \sum_{j=1}^{G} \sum_{i=1}^{n_j} (\mathbf{x}_i - \bar{\mathbf{x}}_j)(\mathbf{x}_i - \bar{\mathbf{x}}_j)^T = \frac{\mathbf{E}}{n-G} \tag{4.0.3}$$

are the *between* and *within* groups covariance matrices, respectively, with mean vector $\bar{\mathbf{x}} = \frac{1}{n}\sum_{i=1}^{n}\mathbf{x}_i$ and group's mean vector $\bar{\mathbf{x}}_j = \frac{1}{n_j}\sum_{i=1}^{n_j}\mathbf{x}_i$.

Fisher's criterion can be translated [Rencher, 2002] to finding the largest eigenpairs $(\lambda, \mathbf{a})$ of the generalised eigenproblem

$$(\mathbf{B} - \lambda \mathbf{W})\mathbf{a} = 0,$$

which can be transformed to the standard eigenproblem

$$(\mathbf{W}^{-1}\mathbf{B} - \lambda I)\mathbf{a} = 0 \tag{4.0.4}$$

or, equivalently

$$(\mathbf{E}^{-1}\mathbf{H} - \tilde{\lambda} I)\mathbf{a} = 0$$

where $\tilde{\lambda} = \frac{G-1}{n-G}\lambda$.

The solution of the eigenproblem (4.0.4) in matrix terms is $\mathbf{BA} = \mathbf{WA\Lambda}$ where $\mathbf{\Lambda}$ is the diagonal matrix of the ordered eigenvalues and $\mathbf{A}$ is the matrix of the corresponding eigenvectors. From the eigenvectors $\mathbf{A}$ of $\mathbf{W}^{-1}\mathbf{B}$ corresponding to $\mathbf{\Lambda}$, we obtain the $N \times s$ - where $s = min(G-1, p)$ - discriminant scores matrix $\mathbf{Z} = \mathbf{XA}$ where

$$\mathbf{X} = \begin{bmatrix} \mathbf{x}_1 \\ \mathbf{x}_2 \\ \vdots \\ \mathbf{x}_n \end{bmatrix}.$$

The matrix $\mathbf{Z}$ shows the s directions of differences among group means, called discriminant functions. The relative importance of each discriminant function can be assessed by considering the corresponding eigenvalue as a proportion of the total. To measure how well the variables separate the groups, we have computed the Roy's statistics η_i^2 that serves as a measure of association for discriminant functions:

$$\eta_i^2 = \frac{\lambda_i}{1+\lambda_i}.$$

The η_i^2 pick the maximum squared correlation between each discriminant function and the best linear combination of the $p-1$ group membership (dummy) variables (see [Rencher, 2002]).

The eigenproblem in (4.0.4) is not symmetric, i.e., $\mathbf{A}$ is of full column-rank and result that $\text{diag}(\mathbf{A}'\mathbf{A}) = \mathbf{I}$. It is usual to normalise the diagonal matrix $\mathbf{A}'\mathbf{W}\mathbf{A}$ such that $\mathbf{A}'\mathbf{W}\mathbf{A} = \mathbf{I}$, i.e:

$$\mathbf{A}_{raw} = \mathbf{A}\text{diag}(\mathbf{A}'\mathbf{W}\mathbf{A})^{-1/2},$$

where $\text{diag}(\mathbf{A}'\mathbf{W}\mathbf{A})^{-1/2}$ is the diagonal matrix containing the elementwise square roots of main diagonal of $(\mathbf{A}'\mathbf{W}\mathbf{A})$. The elements of the normalised matrix $\mathbf{A}_{raw}$ are called *raw coefficients* and the normalization makes the within groups variance of the discriminant scores $\mathbf{Z} = \mathbf{X}\mathbf{A}_{raw}$ equal to 1. The raw coefficients are considered difficult to interpret when we want to evaluate the relative importance of the original variables. One can try to identify the coefficients that are large in magnitude and conclude that the corresponding variables are important for discrimination between the groups. The problem is that such a conclusion is misleading in LDA. The large magnitude may be caused by large between-groups variability, but may also be caused by small within-groups variability [Krzanowsky, 2003]. This problem with the interpretation of raw coefficients can be handled by additional standardization of $\mathbf{A}_{raw}$ which makes all variables comparable:

$$\mathbf{A}_{std} = \text{diag}(\mathbf{W})^{1/2}\mathbf{A}_{raw}.$$

When $p \gg n$, we have the so called *high dimension/small sample size problem*. In this case, the generalised eigenproblem can be ill-posed and the covariance matrix $\mathbf{W}$ is singular. There are different strategies to deal with this problem:

1. **Feature Subset Selection**: in this case, from $\mathbf{x} = [\mathbf{x}_1 \ \mathbf{x}_2 \ \ldots \ \mathbf{x}_p] \in \mathbb{R}^{n \times p}$, we choose an optimal subset of features $\tilde{\mathbf{x}} \in \mathbb{R}^{n \times \tilde{p}}$ with $\tilde{p} < n$;

2. **Dimensionality Reduction**: in this case we map the multidimensional space $\mathbb{R}^p$ into a reduced space $\mathbb{R}^{\tilde{p}}$ such as $\tilde{p} < n$; this is also useful to overcome the singularity of $\mathbf{W}$, i.e., solving the ill-posed generalised eigenproblem by a transformation of the matrix $\mathbf{W}$.

The following sections discuss these topics.

4.1 Feature Subset Selection

When trying to approach to huge database with supervised classification techniques, as in our case, new problems arise in order to identify subsets of variables with high discriminatory power. These problems may be purely mathematical, such as collinearity, indetermination, convergence, etc., or of a statistical nature, particularly overfitting. The use of feature selection algorithms is mainly motivated when classifiers are trained with a limited set of training samples. In these cases, if the number of features is increased, the classification rate of the classifier decreases after a peak [Miller, 2002]. Furthermore, the need to reduce the number of features when designing a recognition system is motivated by computational reasons.

In many classification problems, relevant features are unknown *a priori*. Therefore, many candidate features are introduced to represent the phenomenon better. Unfortunately, it is often true that most of these are either partially or completely redundant to the target. Thus, when the size p of dataset is large, an important primary step in the classification task is to remove the unwanted features.

Feature selection can be considered as an optimization problem: we are asked to select an optimal subset in order to maximise the classification performance of the recognition system. Although it is theoretically possible to consider all subsets and select the best rated one, in practice, this route is mostly unfeasible since the number of feature subsets increases exponentially with the number of features. For example, if we consider our spatial application where 187 features were extracted (cf section 3.3.1), for a maximum feature subset of size 7, a number of 14162×10^{12} feature combinations should be tested.

There are four main ingredients in a typical subset selection method:

a) *Subsets generation procedure*

There are different approaches for facing with the problem of subsets generation, namely: *complete*, *heuristic* and *random generation*. As already remarked, although a complete search is exhaustive, it is often computationally unfeasible. Thus, usually a heuristic search is preferred. This incremental generation procedures starts with no features (or all features) in the optimal subset and in each iteration all remaining features yet to be selected (or eliminated) are considered for selection (or elimination). On the other hand, in the random generation procedures, features are either iteratively added or removed randomly thereafter.

b) *Subsets evaluation function*

To evaluate the goodness of a subset generated from the chosen procedure, the use of an evaluation function is needed. Obviously, an optimal subset is always linked to the evaluation criterion. Generally, an evaluation function tries to measure the capability of a feature, or a subset, to distinguish among different class labels. There are different types of evaluation functions based on:

- *distance*: a feature is preferred to another if the latter induces a smaller difference between the groups.

- *information*: a feature X_1 is preferred to X_2 if the information gain from X_1 is greater than that from X_2, e.g. based on entropy measures.

- *dependence*: a feature is preferred to another if it presents a lower correlation with the groups. This type of evaluation can be viewed as a type distance, but is conceptually different.

- *consistency*: we find out the minimally sized subset that satisfies the acceptable inconsistency rate set by the user. For example, given a training sample S and a subset of features Q we search in the sample for two observation that have the same values for all the features in Q but have conflicting class labels. The subset Q is sufficient to construct hypotesis on consistency if and only if such a pair cannot be found in the training sample [Almuallim and Dietterich, 1994, Dash and Liu, 1997].

- *classifier error*: we find the subset that minimise the misclassification error rate[1].

c) *Stopping criterion*

Without a suitable stopping criterion, the feature selection process may run exhaustively or loop infinity. In general, considering the generation procedure, we can define the

---

[1] This type of evaluation is also classified in the class of *wrapper methods*, meaning that the inductive algorithm is used as the evaluation function. Conversely, the *filter methods* are independent from the inductive algorithm that will finally process the selected subset.

stopping criterion by specifying a predetermined number of selected features, or when a predefinied number of iterations is achieved. Alternatively, we can base the stopping criterion on the evaluation (objective) function, such that, for example, we continue to add (remove) subsequent features until the procedure does not produce any improvement (a better subset).

d) *Validation procedure*

The validation procedure tries to test the validity of the selected subset by a comparison of the results obtained through competing feature selection methods (i.e., changing the generation procedure or the evaluation function or the stopping criterion).

4.1.1 Stepwise Subset Selection

A common suggestion for avoiding the consideration of all subsets is to use stepwise selection. The main advantage of this procedure is that a small, readily interpretable model arises, which contains the most important predictors in a prediction problem. In this section we describe the stepwise selection procedure in detail.

Consider a classical FLDA problem as in (4.0.1) and define the *Wilks Lambda* statistic as

$$\Lambda = \frac{\det(\mathbf{E})}{\det(\mathbf{E} + \mathbf{H})} \qquad (4.1.1)$$

where $\mathbf{E}$ and $\mathbf{H}$ are the within-groups sum of squares and the between-groups sum of squares as in (4.0.3) and (4.0.2), respectively.

Let $\mathbf{X} = [\mathbf{x}_1 \; \mathbf{x}_2 \; \ldots \; \mathbf{x}_p] \in \mathbb{R}^{n \times p}$; then denote also with $\mathbf{Z} = [\mathbf{z}_1 \; \mathbf{z}_2 \; \ldots \; \mathbf{z}_q] \in \mathbb{R}^{n \times q}$, q vectors measured in addition to $\mathbf{X}$. To test the hypothesis that the extra variables in $\mathbf{Z}$ do not contribute anything significant to separating the groups beyond the information already available in $\mathbf{X}$, i.e., whether $\mathbf{Z}$ makes a significant contribution to test $H_0 : \boldsymbol{\mu}_1 = \boldsymbol{\mu}_2 = \cdots = \boldsymbol{\mu}_g$, we calculate [Rencher, 2002]

$$\Lambda(\mathbf{Z}|\mathbf{X}) = \frac{\Lambda([\mathbf{X} \; \mathbf{Z}])}{\Lambda(\mathbf{X})} \qquad (4.1.2)$$

which is distributed as Wilks Lambda $\Lambda_{df_1, df_2, df_3}$ where $df_1 = q$, $df_2 = g - 1$, $df_3 = n - g - p$.

If $q = 1$, i.e., we are interested in the effect of adding a single variable z_i, the (4.1.2) becomes

$$\Lambda(z_i|X) = \frac{\Lambda([X\ z_i])}{\Lambda(X)} \tag{4.1.3}$$

which is distributed as Λ_{1,df_2,df_3}. For $q = 1$ the Λ-statistic in (4.1.3) has an exact F-transformation

$$F = \frac{1 - \Lambda(z_i|X)}{\Lambda(z_i|\ X)} \frac{df_3}{df_2} \tag{4.1.4}$$

which is distributed as F_{df_2,df_3}. The statistic (4.1.3) is often referred to as a partial Λ-statistic; correspondingly, (4.1.4) is called a partial F-statistic. We can rewrite (4.1.3) as

$$\Lambda([X\ z_i]) = \Lambda(z_i|X)\Lambda(X) \leq \Lambda(X)$$

which shows that Wilks Λ can only decrease with an additional variable.

Thus, if there are no variables for which we have *a priori* interest in testing for significance, we can perform a data-directed search for the variables that separates the groups in the best manner. Such a strategy is often called *stepwise subset selection*.

We first describe an algorithm that is usually called *forward selection*:

1. calculate $\Lambda(x_i)$ for each individual variable and choose the one with minimum $\Lambda(x_i)$ (or maximum associated F).

2. calculate $\Lambda(x_i|x_{(1)})$ for each of the $p-1$ variables not entered at step 1, where $x_{(1)}$ indicates the first variable entered and choose the variable with minimum $\Lambda(x_i|x_{(1)})$ (or maximum associated partial F), that is, the variable that adds the maximum separation to the one entered at step 1. Denote the variable entered as $x_{(2)}$.

3. calculate $\Lambda(x_i|[x_{(1)}\ x_{(2)}])$ for each of the $p-2$ remaining variables and choose the one that minimises $\Lambda(x_i|[x_{(1)}\ x_{(2)}])$ (or maximises the associated partial F).

4. Continue this process until the F falls below some predetermined threshold value, say, F_{in}.

A stepwise procedure follows a similar sequence, except that after a variable has entered, the previously selected variables are re-examined to see if each still contributes a significant amount. The variable with the smallest partial F will be removed if the partial F is less than a second threshold value, F_{out}. If F_{out} is the same as F_{in}, there is a very small possibility that the procedure will cycle continuously without stopping. This possibility can be eliminated by using a value of F_{out} slightly less than F_{in}.

Obviously, we can replace the F_{in} and F_{out} with the associated significance levels α_{in} and α_{out}.

The algorithm is as follows:

1. calculate $\Lambda(\mathbf{x}_i)$ for each individual variable and choose the one with minimum $\Lambda(\mathbf{x}_i)$ (or maximum associated F) with $F \geq F_{in}$.

2. calculate $\Lambda(\mathbf{x}_i|\mathbf{x}_{(1)})$ for each of the $p-1$ variables not entered at step 1, where $\mathbf{x}_{(1)}$ indicates the first variable entered and choose the variable with minimum $\Lambda(\mathbf{x}_i|\mathbf{x}_1)$ (or maximum associated partial F) and $F \geq F_{in}$. Denote the variable entered as $\mathbf{x}_{(2)}$.

3. calculate $\Lambda(\mathbf{x}_{(1)}|\mathbf{x}_{(2)})$. If the associated $F \leq F_{out}$ remove $\mathbf{x}_{(1)}$ from the model otherwise keep $\mathbf{x}_{(1)}$.

4. calculate $\Lambda(\mathbf{x}_i|[\mathbf{x}_{(1)}\ \mathbf{x}_{(2)}])$ for each of the $p-2$ remaining variables and choose the one that minimises $\Lambda(\mathbf{x}_i|[\mathbf{x}_{(1)}\ \mathbf{x}_{(2)}])$ (or maximises the associated partial F) with $F \geq F_{in}$. Denote the variable entered as $\mathbf{x}_{(3)}$.

5. calculate $\Lambda(\mathbf{x}_{(1)}|[\mathbf{x}_{(2)}\ \mathbf{x}_{(3)}])$. If the associated $F \leq F_{out}$ remove $\mathbf{x}_{(1)}$ from the model otherwise $\mathbf{x}_{(1)}$ keep. Also, calculate $\Lambda(\mathbf{x}_{(2)}|[\mathbf{x}_{(1)}\ \mathbf{x}_{(3)}])$. If the associated $F \leq F_{out}$ remove $\mathbf{x}_{(2)}$ from the model otherwise keep $\mathbf{x}_{(2)}$.

6. continue this process from step 4 until there are no other variables for which $F \geq F_{in}$.

4.1.2 Stepwise Subset Selection: Sensitivity Analisys

When we perform LDA with a high dimensional dataset using statistical computer packages, it is common to obtain a subset of variables from a stepwise procedure and then evaluate its discriminant power by the *leave one out cross validation* (LOCV) [Stone, 1974, Hastie et al., 2009]. This procedure involves to leave out a single observation from the original sample to use as a validation data, and the remaining observations as the training data from which to classify the one left out; this is repeated in such a way that each observation in the sample is used once as the validation data. The resulting misclassification error rate is also called LOCV error rate.

A defect of the stepwise procedure is that there is no guarantee that the locally stable subset at which the routine terminates is in fact the optimum [Hawkins, 1976]. Moreover, stepwise tend to yield results that are not replicable [Thompson, 1995, 2001] because the procedures are sample-specific. In fact, small differences that are specific to the sample may give one predictor an advantage over another that it would not have seen in future samples. This error is compounded by the fact that a sequence of decisions is made based on that erroneous decision, potentially corrupting the subsequent choices, making stepwise sample results unlikely to generalise.

For these reasons, we decided to study the sensitivity of the stepwise procedure to changes in the sample units with respect to the LOCV error rate. In particular, the Figures 4.1 shows the APER and LOCV error rate registered for different significance levels in the spatial domain and the corresponding subset cardinalities identified by the stepwise procedure. We observe that the stepwise procedure shows very different results corresponding

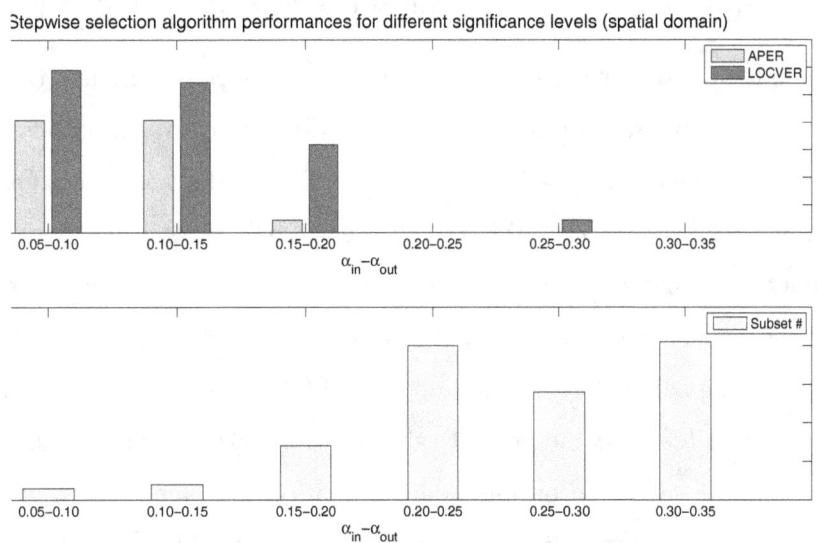

Figure 4.1: Stepwise performance for different significance levels (spatial domain)

to different significance levels. Moreover, increasing the significance levels the procedure selects an increasing number of variables and this seems to reflect a reduction in the misclassification error. However, the LOCV error rate distributions show that this is not true (see Figure 4.2 on the facing page). In fact, for fixed significance levels, by performing a jackknife procedure [Quenouille, 1949, 1956, Miller, 1974] and registering, in turn, the classification performances of the stepwise subset selection procedure, we obtain the distributions of the LOCV error rate for each possible subset cardinality; we observe that the performance of the stepwise procedure, both in terms of LOCV error rate and number of selected variables, is very sensitive to variations in the sample units.

Similarly, the Figure 4.3 on page 55 show the APER and LOCV error rate registered for different significance levels in the spatio-temporal domain and the corresponding subset cardinalities identified by the stepwise procedure.

We observe that the performance of the stepwise procedure in terms of LOCV error rate does not improve with increasing levels of significance, i.e. when the number of selected variables increases. Moreover, the Figure 4.4 on page 56 shows that. Again, we observe that the performance of the stepwise procedure, both in terms of LOCV error rate and number of selected variables, is sample-specific.

4.1.3 Improving Stepwise Subset Selection: the JSS+E Algorithm

Although the stepwise selection is affected by a series of issues, this procedure is one of the most widely used statistical subset selection procedure. Methods that use the jackknife procedure [Quenouille, 1949, 1956, Miller, 1974] in the selection steps are also useful. For example, *Schemata Search* [Moore and Lee, 1994] starts either from the empty set or the complete set, and at each iteration it finds the best subset by either removing or adding only one feature from the subset. It evaluates each subset using LOCV by selecting, at each iteration, the subset having the *minimum* LOCV error rate. Other types of methods are the so-called *best-first* methods [Dash and Liu, 1997]. For example, at each step, we take only the best subset according to an evaluation function. Thus, we consider all possible subsets produced by adding only one feature to it. The procedure continues by ranking the produced subsets and taking the new best-first ranked one. We may impose a stopping

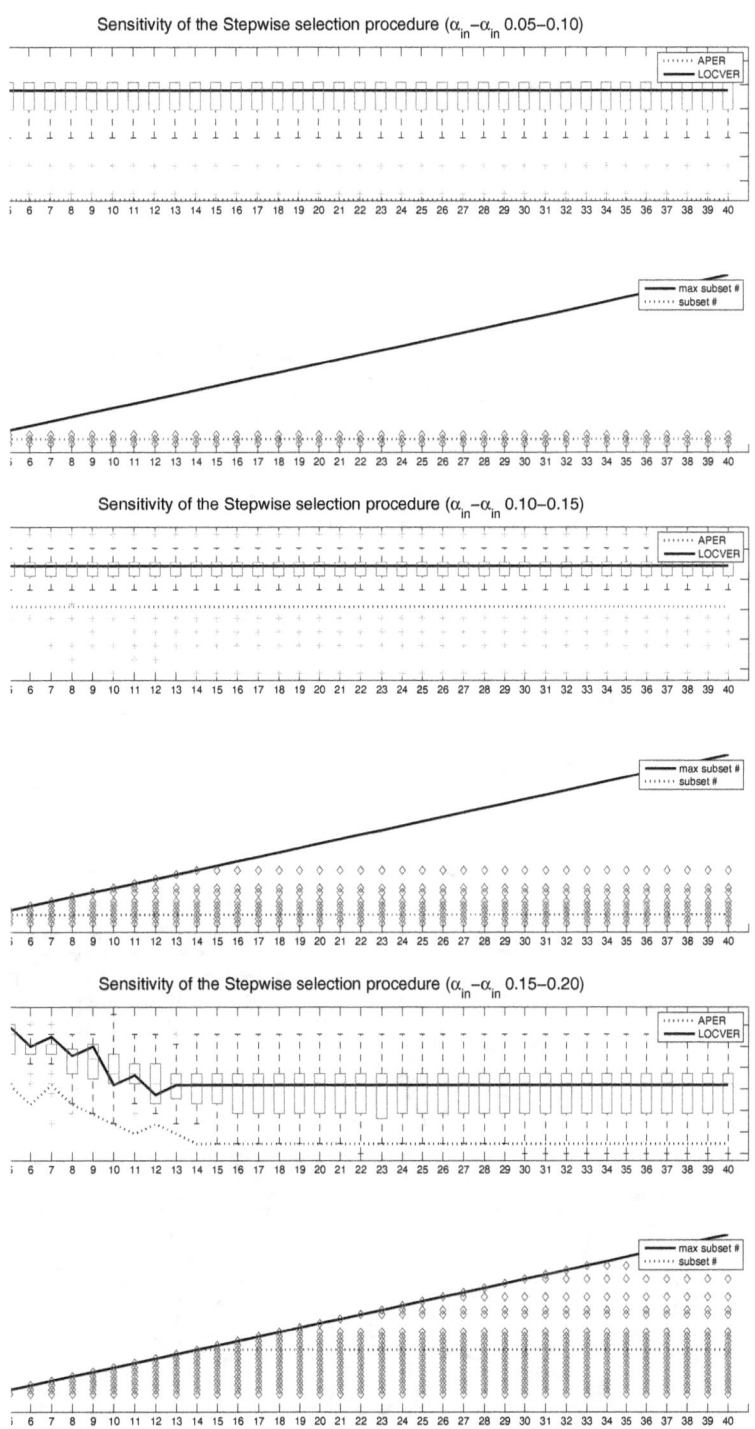

Figure 4.2: LOCV error rate distributions for different subset cardinalities (jackknified procedure, spatial domain)

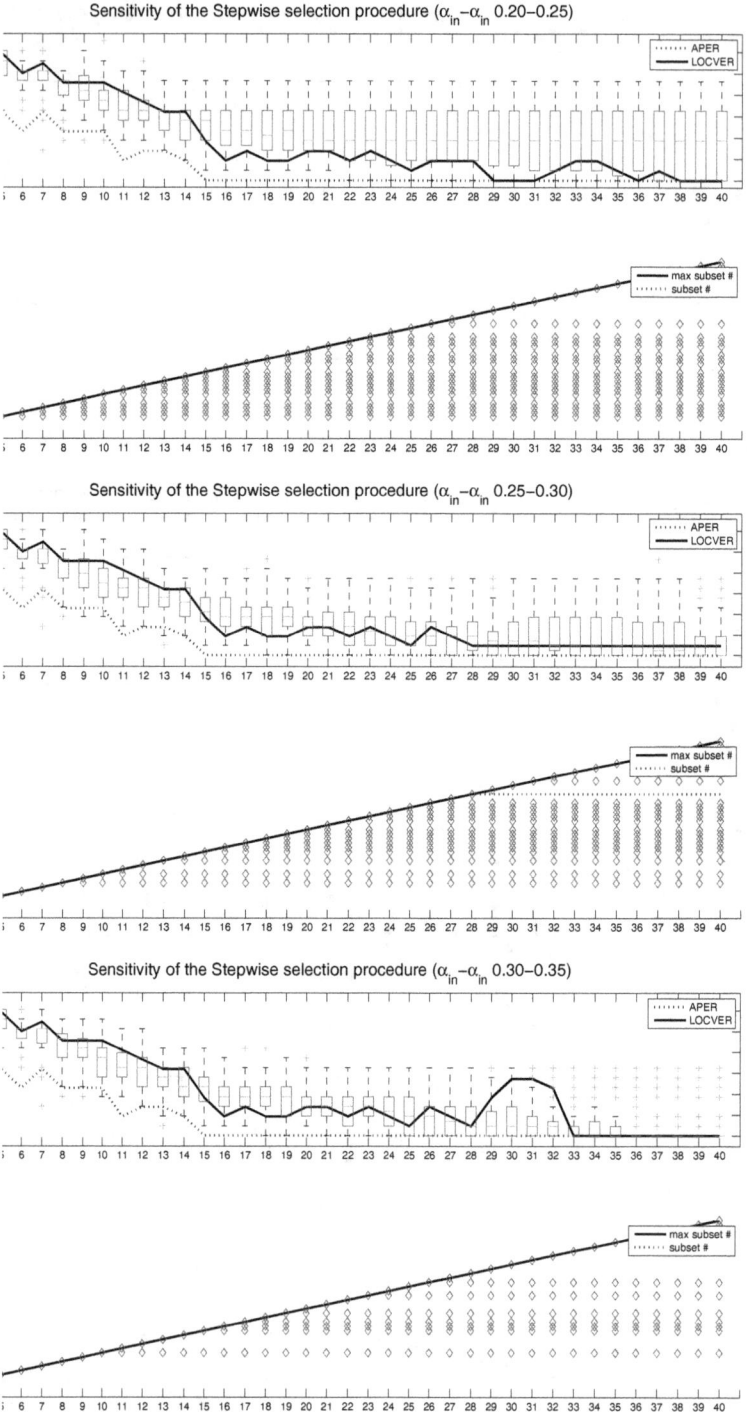

Figure 4.2 (continue)

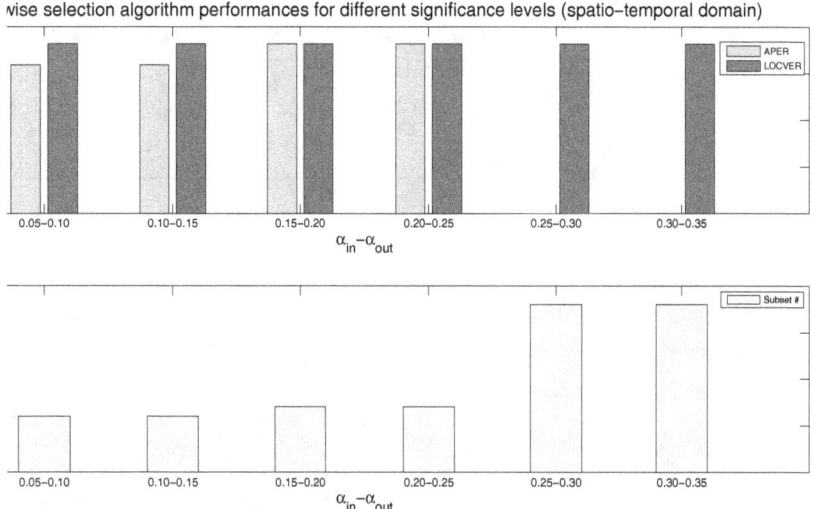

Figure 4.3: Features ranking in spatial domain

rule (e.g. by limiting the size of the ranked queue) or loop the iteration without a stopping criterion. In this latter case the procedure is equivalent to that of an exhaustive search.

In this work we propose to combine the stepwise procedure with the exhaustive search. Moreover, since the stepwise subset selection procedure is sample-specific, the variable selection phase is also subject to a jackknife procedure as recommended by White and Liu [White and Liu, 1993] and by Hastie *et alii* [Hastie et al., 2009]. Omitting each subject in turn, this procedure may provide a way of reducing bias of error rate estimation for the classifier.

Thus, specifcally, we perform the following algorithm:

1. Inizialise $j = 1$;

2. leave out the j-th unit from the data matrix;

3. perform the stepwise selection procedure and register the selected features;

4. increase j to $j = j + 1$ and repeat steps 2 and 3 until $j = n$ (n is the number of units in the original sample);

5. rank the selected features (from the most selected to the less selected) and choose the first k ranked features;

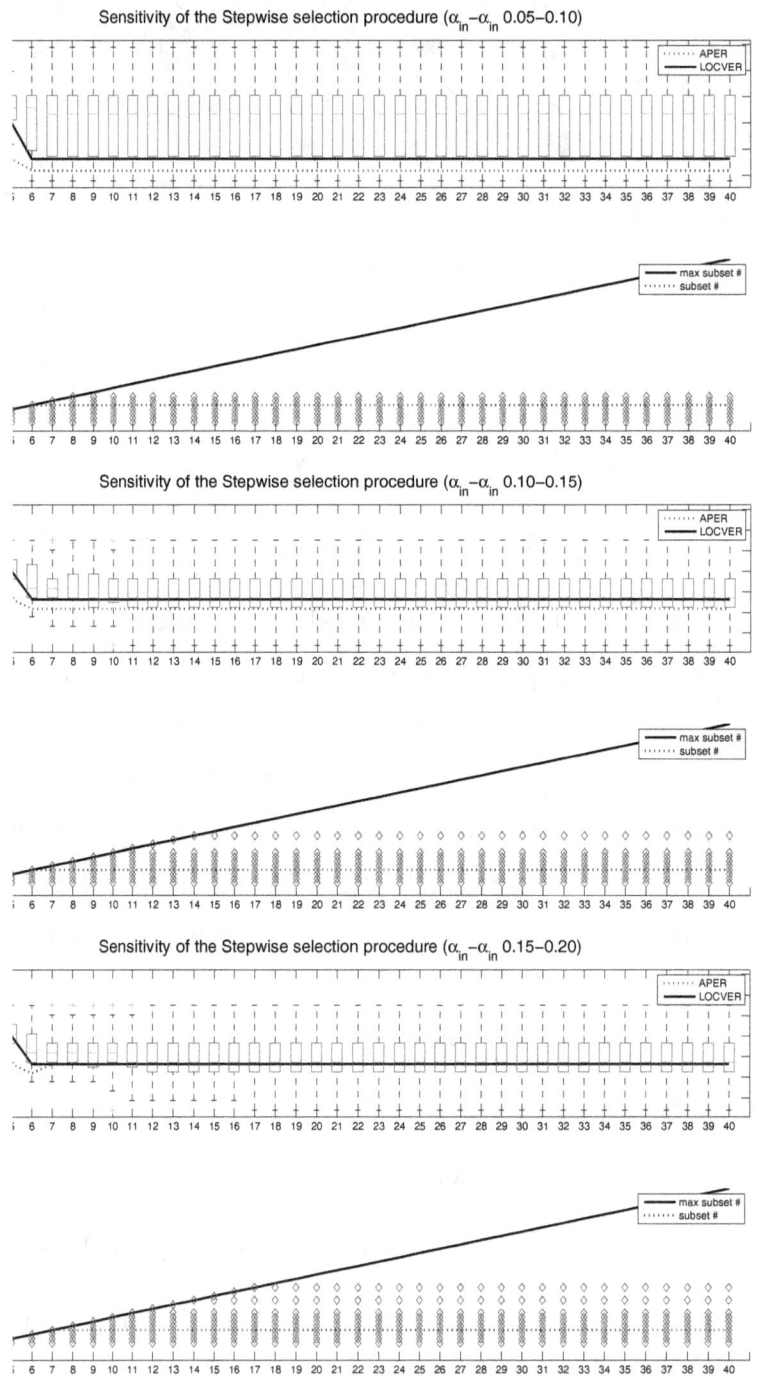

Figure 4.4: LOCV error rate distributions for different subset cardinalities (jackknified procedure, spatio-temporal domain)

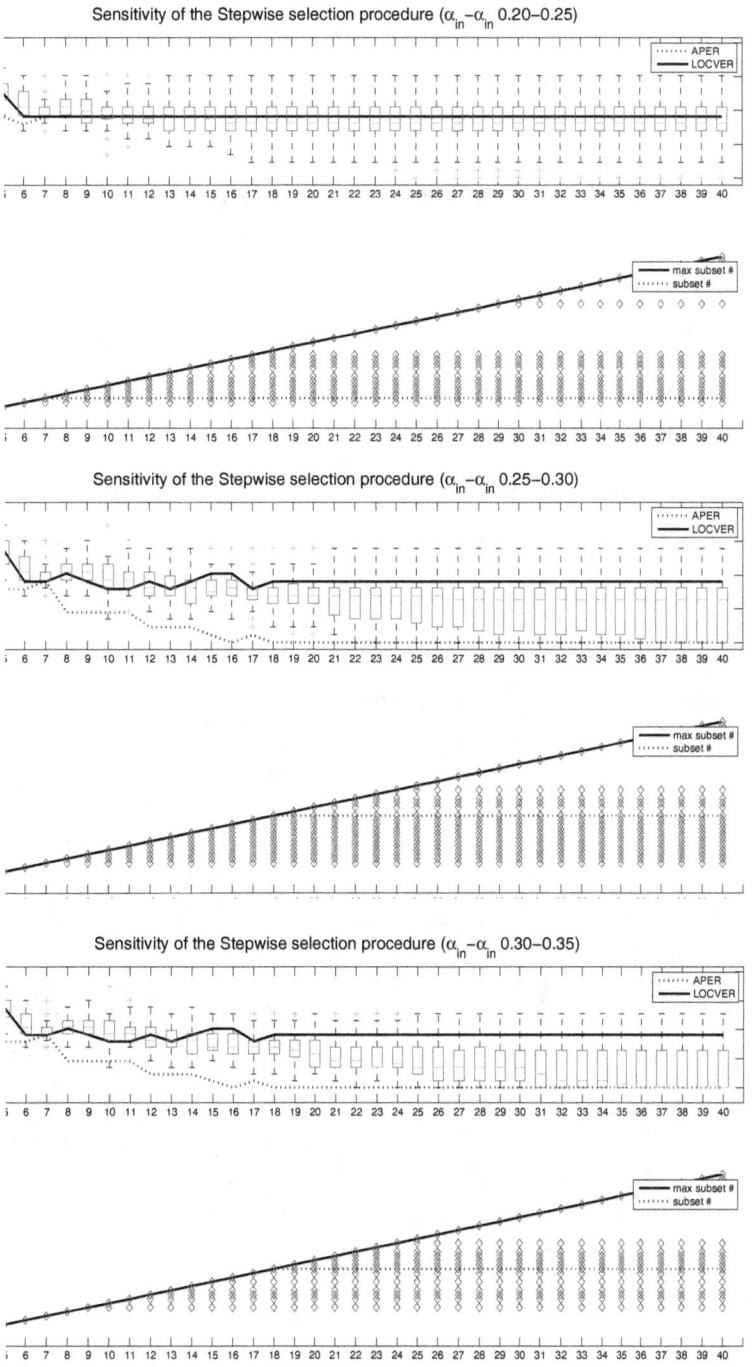

Figure 4.4 (continue)

58

6. perform an exhaustive selection on the first k ranked features evaluating each subset using LOCV error rate.

We call steps 1-6 Jackknified Stepwise Selection + Exhaustive search (JSS+E).

4.2 Classifying Raynaud's Phenomenon

In a prognostic modelling study, Steyerberg et al. [Steyerberg et al., 2000] found that stepwise selection with a low α_{in} (0.05) led to a relatively poor model performance and indicates to consider higher levels of alpha. Following these suggestions, we first process steps 1-5 of JSS+E for different levels of α_{in} and α_{out} in the interval $(0.05, 0.35)$.

Considering our classification problem in terms of the spatial domain approach, the extracted features were enumerated from 1 to 182 (F.id) according to section 3.3.1; the following Figure 4.5 gives a pictorial representation of the number $n_{F.id}$ of times each variable is selected in the JSS+E procedure.

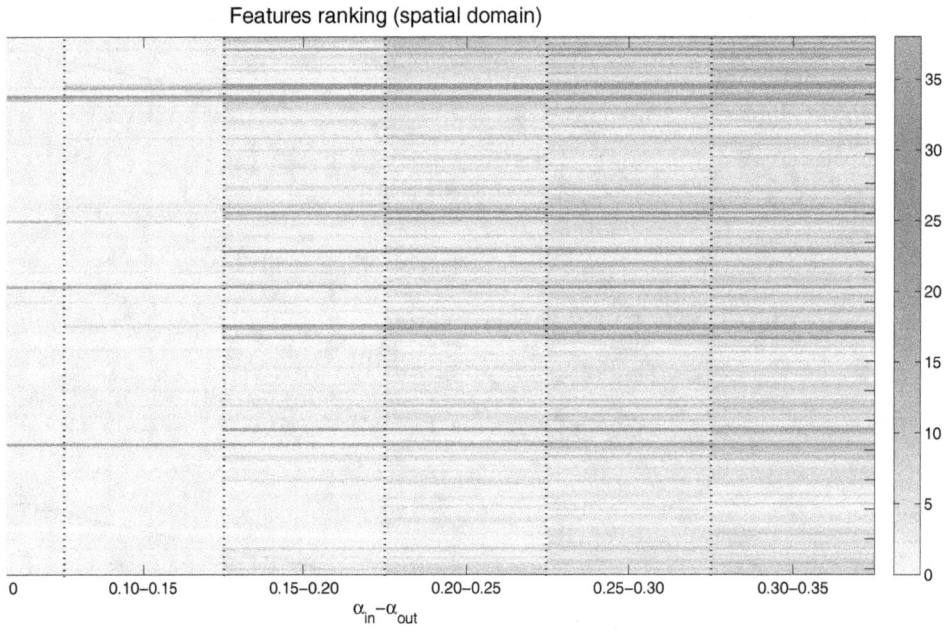

Figure 4.5: Features ranking in spatial domain

It is easy to observe that for low significance levels $(0.05 - 0.10)$, only 3 varibles were frequently selected. When raising levels of α_{in} and α_{out}, an increasing number of variables were selected by the procedure. In particular, the left hand conditional variance $\hat{v}_2^2$ of the GMRF (F.id 21), the contrast $T_1(0,5)$ in (3.2.6) (F.id 85) and the energy $T_3(8,-8)$ in (3.2.8) (F.id 138) were always selected. By increasing the significance levels beyond the interval $0.15 - 0.20$, are also selected the vertical right hand spatial GMRF parameter $\hat{\beta}_{(0,1)1}$ (F.id 18) together with the horizontal right hand spatial GMRF interaction parameter $\hat{\beta}_{(1,0)1}$ (F.id 17), the energy $T_3(0,8)$ (F.id 98) in (3.2.8) and the right hand conditional variance $\hat{v}_1^2$ (F.id 22) of the GMRF, are also selected.

For the sake of simplicity, Table 4.1 shows the 20 most selected features by the first 5 steps of the JSS+E procedure.

α_{in}	0,05	α_{in}	0,1	α_{in}	0,15	α_{in}	0,2	α_{in}	0,25	α_{in}	0,3
α_{out}	0,1	α_{out}	0,15	α_{out}	0,2	α_{out}	0,25	α_{out}	0,3	α_{out}	0,35
F.id	$n_{F.id}$	F.id	$n_{F.id}$	F.id	$n_{F.id}$	F.id	$n_{F.id}$	F.id	$n_{F.id}$	F.id	$n_{F.id}$
85	37	85	38	18	33	98	34	18	35	18	33
21	29	138	34	85	33	22	32	98	34	98	33
138	29	21	30	98	33	18	31	22	30	22	32
63	14	18	16	17	29	85	31	17	28	85	30
22	12	22	12	73	27	17	30	60	28	17	29
90	3	17	6	22	24	60	28	85	28	5	24
18	2	63	5	138	24	73	27	73	27	3	23
80	2	98	5	21	22	138	25	5	26	60	23
3	1	58	3	60	21	21	21	138	25	73	23
5	1	2	2	5	18	62	21	62	24	102	22
58	1	5	2	133	18	5	20	133	21	99	21
100	1	24	2	102	17	133	19	1	20	133	21
125	1	35	2	62	13	1	18	21	20	138	21
1	0	60	2	99	12	99	18	99	19	7	20
2	0	80	2	1	10	77	16	102	19	62	20
4	0	90	2	2	10	102	14	7	18	1	19
6	0	120	2	77	10	58	12	77	16	21	17
7	0	139	2	58	9	2	11	3	15	58	16
8	0	3	1	3	7	3	11	52	14	12	15
9	0	10	1	23	6	7	10	35	13	16	13

Table 4.1: Frequency distribution (spatial domain) of the first 20 most selected features for different $\alpha_{in} - \alpha_{out}$

Similarly, in the space-time domain approach, the extracted features were enumerated from 1 to 279 (F.id) according to section 3.3.2; the following Figure 4.6 illustrates the importance of each variable in terms of selections in the JSS+E procedure. In this case, many fetaures are frequently selected even with low significance levels. In particular, the first 5 steps of the JSS+E algorithm often select the contrast $T_1(6,-6,0)$ (F.id 127) in (3.2.6) together with the diagonal south-east GMRF interaction parameter (F.id 17) $\beta_{(1,-1,0)2}$ (see

Table 3.5) referred to the left hand and the energy $T_3(6,0,1)$ (F.id 185) in (3.2.8).

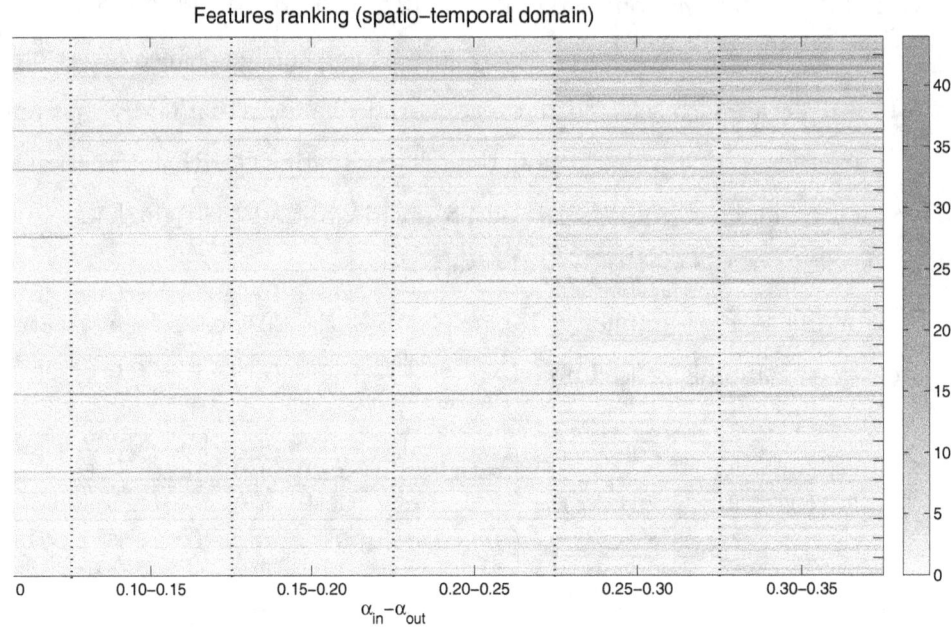

Figure 4.6: Features ranking in spatio-temporal domain

In order to perform the step 6 of the JSS+E algorithm, we consider the first $k = 20$ ranked features. This choice depends on the computational time required to perform the exhaustive search, since the number of combinations increases esponentially with the number of considered features. In particular, for $k = 20$ we have 1048555 possible subsets[2]. For each subset, the LOCV error rate was computed and used (as evaluation function) to evaluate the goodness of the subset. Figure 4.7 on page 63 shows the *minimum* LOCV error rate registered in the spatial domain approach and computed for each subset cardinality $k = 2, 3, \ldots, 20$ and for different levels of α_{in} and α_{out}. We remark here that for levels of $\alpha_{in} = 0.05$ and $\alpha_{out} = 0.10$, the steps 1-5 of the JSS+E procedure select 13 features.

We observe also that for $k > 15$, the LOCV error rate increases, probably by overfitting[3]. Furthermore, even if it is not required mathematically, to have more stable discriminant functions, ideally the smallest n_j (13 in our case) should exceed the number of variables, p

---

[2] The procedure was processed with a intel centrino 2 Duo 2,40 GHz and the algorithm was elapsed in 44915,30 seconds.

[3] Increasing the subset cardinality, the APER still remains close to 0 while the prediction capabilities of the model are worse and worse.

| α_{in} | 0,05 | α_{in} | 0,1 | α_{in} | 0,15 | α_{in} | 0,2 | α_{in} | 0,25 | α_{in} | 0,3 |
α_{out}	0,1	α_{out}	0,15	α_{out}	0,2	α_{out}	0,25	α_{out}	0,3	α_{out}	0,35
F.id	$n_{F.id}$	F.id	$n_{F.id}$	F.id	$n_{F.id}$	F.id	$n_{F.id}$	F.id	$n_{F.id}$	F.id	$n_{F.id}$
127	44	127	44	127	44	127	42	17	36	17	37
104	19	17	29	17	34	17	38	33	30	33	29
17	18	185	26	185	27	185	30	185	30	1	28
18	17	69	23	69	24	69	27	127	28	185	28
69	16	49	22	49	22	49	26	49	24	69	26
185	16	225	19	225	20	33	22	1	23	49	25
49	15	18	17	18	16	225	20	18	22	127	25
225	12	33	6	33	14	21	18	69	22	18	22
102	4	102	6	21	13	18	14	23	19	23	21
229	4	104	6	102	9	1	12	12	17	11	20
189	3	54	5	54	5	12	11	225	17	12	19
14	2	14	4	1	4	14	7	21	15	21	17
170	2	189	4	12	4	23	7	14	14	225	16
9	1	229	4	14	4	264	7	8	13	96	15
10	1	12	3	104	4	138	6	11	13	8	14
12	1	16	3	189	4	8	5	13	12	13	14
22	1	21	3	264	4	9	5	96	12	14	14
33	1	165	3	16	3	54	5	29	10	28	12
43	1	22	2	23	3	165	5	31	10	44	12
65	1	65	2	165	3	275	5	44	10	4	11

Table 4.2: Frequency distribution (spatio-temporal domain) of the 20 most selected features for different $\alpha_{in} - \alpha_{out}$

[Rencher, 2002]. In general, the significance levels that seems to show the best performances in terms of LOCV error rate correspond to 0,15-0,20. For classification purpose, we choose to select the subsets with *minimum* LOCV error rates and, among these, the one with *minimum* cardinality. This corresponds to the subset with cardinality 11 and $\alpha_{in} - \alpha_{out}$ equal to $0.15 - 0.20$. The following Table 4.3 on the following page shows the results of the JSS+E procedure in the spatial domain approach.

Table 4.3 also reports the (modified) Borda count; for each subset cardinality, we assign 1 point to the column which shows the *minimum* LOCV error rate and 0 otherwise. Computing the sum of awarded points, we have an indication of the performance, in terms of LOCV error rate, of each significance level. Specifically, the significance levels 0.15-0.20 (column c) and 0.25-0.30 (column e) seem to show the best performances. Finally, for subset cardinalities in the interval [8,16] and for significance levels more then 0.15-0.20, the LOCV error rates are beyond $11,4\%$.

The following Figure 4.8 on page 63 depict the LOCV error rate calculated in the spatio-temporal domain approach, for subset cardinalities 2-20 and for different significance levels.

# subset	a α_{in}: 0.05 α_{out}: 0.10 LOCVER	b α_{in}: 0.10 α_{out}: 0.15 LOCVER	c α_{in}: 0.15 α_{out}: 0.20 LOCVER	d α_{in}: 0.20 α_{out}: 0.25 LOCVER	e α_{in}: 0.25 α_{out}: 0.30 LOCVER	f α_{in}: 0.30 α_{out}: 0.35 LOCVER	min	Borda count a	b	c	d	e	f
2	0,250	0,250	0,273	0,273	0,273	0,273	0,250	1	1	0	0	0	0
3	0,182	0,205	0,205	0,205	0,250	0,205	0,182	1	0	0	0	0	0
4	0,182	0,182	0,182	0,182	0,182	0,182	0,182	1	1	1	1	1	1
5	0,205	0,182	0,182	0,182	0,182	0,182	0,182	0	1	1	1	1	1
6	0,205	0,182	0,159	0,159	0,182	0,182	0,159	0	0	1	1	0	0
7	0,205	0,159	0,114	0,159	0,159	0,159	0,114	0	0	1	0	0	0
8	0,205	0,159	0,114	0,114	0,114	0,114	0,114	0	0	1	1	1	1
9	0,205	0,136	0,091	0,091	0,091	0,091	0,091	0	0	1	1	1	1
10	0,205	0,136	0,068	0,068	0,068	0,068	0,068	0	0	1	1	1	1
11	0,250	0,136	0,045	0,045	0,068	0,045	0,045	0	0	1	1	0	1
12	0,295	0,136	0,045	0,045	0,068	0,045	0,045	0	0	1	1	0	1
13	0,386	0,136	0,045	0,068	0,068	0,068	0,045	0	0	1	0	0	0
14		0,114	0,068	0,091	0,068	0,091	0,068	0	0	1	0	1	0
15		0,159	0,068	0,091	0,068	0,091	0,068	0	0	1	0	1	0
16		0,159	0,068	0,114	0,068	0,114	0,068	0	0	1	0	1	0
17		0,182	0,091	0,136	0,091	0,136	0,091	0	0	1	0	1	0
18		0,182	0,136	0,159	0,114	0,159	0,114	0	0	0	0	1	0
19		0,227	0,182	0,205	0,182	0,182	0,182	0	0	0	0	1	1
20		0,295	0,227	0,318	0,182	0,205	0,182	0	0	0	0	1	0
min	0,182	0,114	0,045	0,045	0,068	0,045	0,045	3	3	13	8	12	7

Table 4.3: Results of the JSS+E procedure (spatial domain)

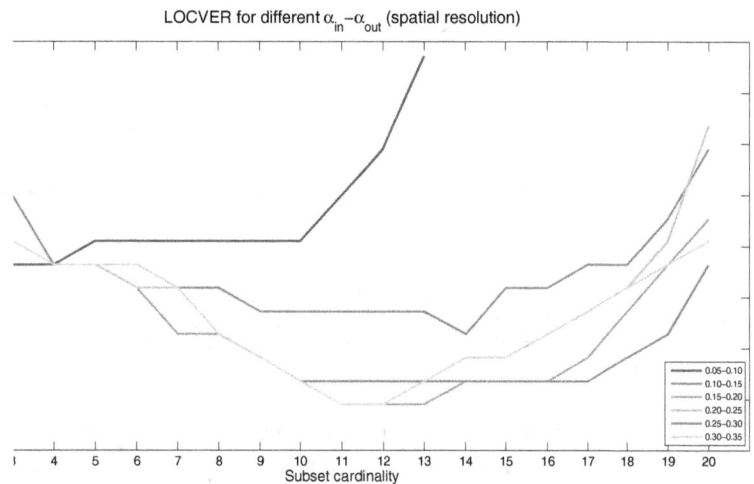

Figure 4.7: LOCVER for different significance levels (spatial domain)

The patterns show that for subset cardinalities more then 13, the LOCV error rate increases.

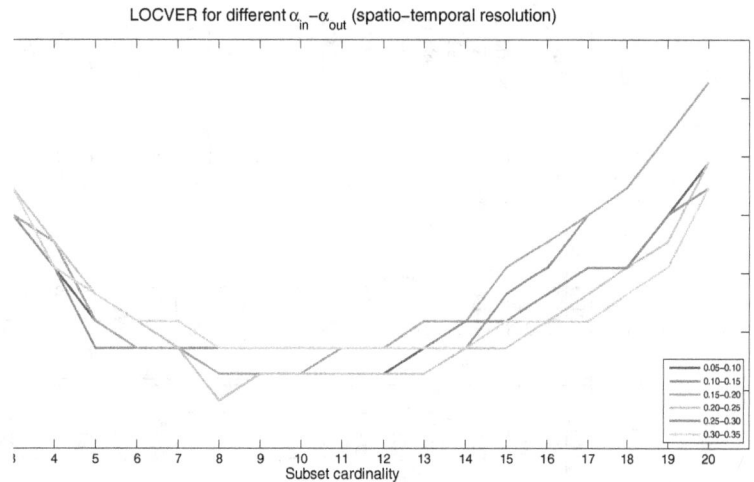

Figure 4.8: LOCVER for different significance levels (spatio-temporal domain)

Moreover, we found that for subsets cardinalities in the interval $[7, 13]$, the LOCV error rate seems to be very conservative for different significance levels (cf Table 4.4 on the following page). The (modified) Borda count shows that the best performances in terms of LOCV error rate are registered for $\alpha_{in} - \alpha_{out}$ equal to $0.05 - 0.10$ and $0.20 - 0.25$. Thus, for the classification purpose in the spatio-temporal domain, we choose to select the subsets with *minimum* LOCV error rates and, among these, the one with *minimum* cardinality. This

# subset	a α_{in}: 0.05 α_{out}: 0.10 LOCVER	b α_{in}: 0.10 α_{out}: 0.15 LOCVER	c α_{in}: 0.15 α_{out}: 0.20 LOCVER	d α_{in}: 0.20 α_{out}: 0.25 LOCVER	e α_{in}: 0.25 α_{out}: 0.30 LOCVER	f α_{in}: 0.30 α_{out}: 0.35 LOCVER	min	Borda count a	b	c	d	e	f
2	0,295	0,318	0,318	0,341	0,318	0,318	0,295	1	0	0	0	0	0
3	0,250	0,250	0,250	0,273	0,273	0,273	0,250	1	1	1	0	0	0
4	0,205	0,205	0,227	0,227	0,205	0,205	0,205	1	1	0	0	1	1
5	0,159	0,136	0,159	0,182	0,182	0,182	0,136	0	1	0	0	0	0
6	0,136	0,136	0,136	0,159	0,159	0,159	0,136	1	1	1	0	0	0
7	0,136	0,136	0,136	0,136	0,159	0,159	0,136	1	1	1	1	0	0
8	0,114	0,136	0,114	0,091	0,136	0,136	0,091	0	0	0	1	0	0
9	0,114	0,136	0,114	0,114	0,136	0,136	0,114	1	0	1	1	0	0
10	0,114	0,136	0,114	0,114	0,136	0,136	0,114	1	0	1	1	0	0
11	0,114	0,136	0,136	0,114	0,136	0,136	0,114	1	0	0	1	0	0
12	0,114	0,136	0,136	0,114	0,136	0,136	0,114	1	0	0	1	0	0
13	0,136	0,136	0,159	0,136	0,159	0,136	0,136	1	1	0	1	0	1
14	0,136	0,136	0,159	0,136	0,159	0,159	0,136	1	1	0	1	0	0
15	0,159	0,182	0,205	0,159	0,182	0,159	0,159	1	0	0	1	0	1
16	0,159	0,205	0,227	0,136	0,159	0,159	0,136	0	0	0	1	0	0
17	0,182	0,250	0,250	0,182	0,205	0,182	0,159	0	0	0	0	0	0
18	0,205	0,273	0,273	0,205	0,205	0,205	0,182	0	0	0	0	0	1
19	0,250	0,318	0,318	0,227	0,250	0,205	0,205	0	0	0	0	0	1
20	0,295	0,364	0,364	0,295	0,273	0,273	0,273	0	0	0	0	1	1
min	0,114	0,136	0,114	0,091	0,136	0,136	0,091	11	6	5	10	2	7

Table 4.4: Results of the JSS+E procedure (spatio-temporal domain)

corresponds to the subset with cardinality 8 and levels of $\alpha_{in} - \alpha_{out}$ equal to $0.20 - 0.25$.

4.2.1 FLDA in the Spatial Domain Approach

Enumerating the features from 1 to 182 (F.id) according to the section 3.3.1, Table 4.5 shows the variables selected by the JSS+E algorithm.

F.id	Feature	Description
5	$\hat{b}_{51}$	GMRF trend paramter related to s_x^2 of the design matrix D_1 (cf 3.3.1) (right hand)
17	$\hat{\beta}_{(1,0)1}$	GMRF spatial interaction parameter for the horizontal direction (right hand)
18	$\hat{\beta}_{(0,1)1}$	GMRF spatial interaction parameter for the vertical direction (right hand)
21	$\hat{v}_2^2$	conditional variance estimation (left hand)
22	$\hat{v}_1^2$	conditional variance estimation (right hand)
58	$T_3(8,0)$	Energy, horizontal direction lag 8
62	$T_2(8,0)$	Homogeneity, horizontal direction lag 8
73	$T_3(0,3)$	Energy, vertical direction lag 3
85	$T_1(0,5)$	Contrast, vertical direction lag 5
98	$T_3(0,8)$	Energy, vertical direction lag 8
99	$T_5(0,8)$	Mutual Information, vertical direction lag 8

Table 4.5: JSS+E algorithm best subset (JSS+E, $\alpha_{in} - \alpha_{out}$ 0.15-0.20, subset cardinality 11)

This subset of selected features consists of a total of 11 discriminant variables mainly related to the vertical and horizontal directions. Specifically, 5 of the selected variables are related to the parameters of the GMRF while the remaining ones are represented by the indices T_1, T_2, T_3 and T_5. Variables related to the vertical direction are defined for spatial lags ranging from 3 to 8, while variables related to the horizontal one are related to lag 8.The presence of scleroderma leads to a progressive destruction of the microvasculature from distal to proximal sections, thus explaining the differences observed in the spatial lags. The wider spatial lag observed for the horizontal direction probably captures information about the orderliness between fingers; the onset of thermoregulatory processes occur mainly both in the nail-bed (before) and in high phalanges (after), with more thermal exchange between the vessels and adjacent tissues. Then, the thermal differences between the different tissues arise by emphasizing the horizontal component. The wider spatial lags observed for the vertical direction could be likely linked to the specific geometry of the finger vasculature and the expression of the functional impairment secondary to the disease; in fact, larger finger vessels run longitudinally and parallelously, while only a few artero-venous shunts run transversally. Also, gradients are most noticeable in the vertical direction corresponding to low spatial lags, because it follows the normal temperature drop associated with the proximal-distal direction.

Performing an LDA procedure on the selected subset shown in Table 4.5, we first test the statistical significance of differences between means in different groups. The following Table 4.6 reports the Wilk's Lambda statistics; the p-values show significance of the test, meaning that the differences within the group mean discriminant scores are greater than what could be attributed to sampling error. The first eigenvalue accounts for a substantial proportion of the total. In fact, the importance of the first root is 0,74; thus the mean vectors lie largely in the first dimension. The proportion of the second eigenvalue on the total is 0,26. The squared canonical correlations between each of the two discriminant functions and the grouping (dummy) variables are 0,803 and 0,589, respectively.

Eigenvalues	C.V. importance	Roy's statistic	Wilk's Lambda	F	P-value
4,096	0,740	0,803	0,081	7,113	0,000
1,437	0,260	0,589	0,410	4,598	0,000

Table 4.6: Wilk's Lambda (JSS+E, $\alpha_{in} - \alpha_{out}$ 0.15-0.20, subset cardinality 11)

The Figure 4.9 depicts the canonical variables produced by the model; the subjects seem to clusterise in 3 groups.

The confusion matrix resulting from LDA gives a 4.5% estimate of the apparent error rate and, performing the Leave-One-Out Cross Validation procedure, the estimated LOCV error rate is still 4.5%. Specifically, Table 4.7 shows the confusion matrix computed for the training set and for the leave-one-out cross-validation procedure. It can be seen that, 2 SSc are wrongly classified as PRP. These correspond to the individuals (id) 36 and 43 (cf Table 4.9) for which we have additional useful clinical informations to explain the misclassifications. In particular:

- subject (id) 36 is a longstanding diffuse SSc for which the vascular damage is well advanced. In this condition, the cutaneous thermoregulation is systematically damaged and no vaso-active reaction can be expressed. Therefore, only passive thermal exchange with the surrounding warmer environment can help into re-establishing basal condition, as the PRP case. By the way, in this case, early assessment through basal condition is not significant;

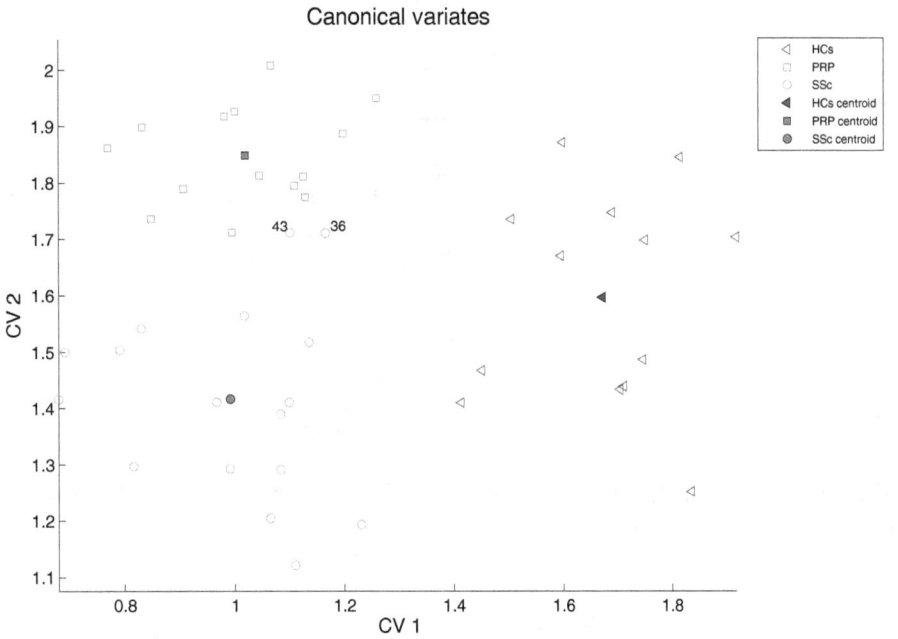

Figure 4.9: Canonical variates (JSS+E, $\alpha_{in} - \alpha_{out}$ 0.15-0.20, subset cardinality 11)

- Subject (id) 43 suffers for Sjögren syndrom for which no studies about the fingers thermoregulation process are available to establish possible differencies with SSc or PRP features. However, this syndrom resembles closely both clinical, autoimmune, haematic features of SSc, with particular reference to the vasomotor tone and vascular impairment; while the subject is classified as SSc due to morphological aspects, functional aspects typical of a PRP occurs.

		Classified		
		HCs	PRP	SSc
Training set	HCs	13	0	0
	PRP	0	14	0
	SSc	0	2	15
	APER	0,0454		
Cross-validation	HCs	13	0	0
	PRP	0	14	0
	SSc	0	2	15
	LOCVER	0,0454		

Table 4.7: Confusion Matrix (JSS+E, $\alpha_{in} - \alpha_{out}$ 0.15-0.20, subset cardinality 11)

Table 4.8 shows the estimated standardised coefficients of the discriminant functions which provide information on the relative impact of the predictors on the dependent variables.

Feature id	Functions	
	1	2
5	0,19	0,51
17	1,39	0,24
18	-0,39	-0,23
21	0,61	-0,49
22	1,33	1,19
58	-3,48	2,05
62	2,21	0,11
73	-16,66	-5,55
85	-1,83	0,00
98	17,18	3,39
99	-2,09	-1,83

Table 4.8: Standardised coefficients for the two discriminant functions (JSS+E, $\alpha_{in} - \alpha_{out}$ 0.15-0.20, subset cardinality 11)

We observe that the variables with F.id 98, 73, 58, 99, 22 contribute most to separating the groups in the two discriminant functions. Furthermore, the variables 62, 85, 17 have a certain importance with respect to the first function. In particular, we remark that the higher (in magnitude) standardised coefficients corresponds to the variables with F.id 98 and 73 which are related to the vertical direction. In fact, in basal condition, the cutaneous temperature distribution is largely affected by the thermal properties of the underlying anatomy; thus, the temperature gradient is largely dominated by the thermal track of longitudinal vessels.

The Table 4.9 on the next page lists the subjects (id) and the corresponding group prediction both considering the training set and performing a leave-one-out cross-validation procedure.

To get more insight in the misclassifications, we plot the rewarming curves (Figure 4.10) of the misclassified subjects both respect to the orignal and to the predicted group. According to the section 2.1, the rewarming curves represent the objects $(t, \bar{u})$ on a coordinates system where $\bar{u} = \frac{1}{10} \sum_{h=1}^{2} \sum_{f=1}^{5} u_{hf}$. We observe that the rewarming curves, also, seem to indicate a different clusterization for the misclassified subjects and to confirm the predicted groups.

id	Group	Classified	Cross-validate
1	1	1	1
2	1	1	1
3	1	1	1
4	1	1	1
5	1	1	1
6	1	1	1
7	1	1	1
8	1	1	1
9	1	1	1
10	1	1	1
11	1	1	1
12	1	1	1
13	1	1	1
14	2	2	2
15	2	2	2
16	2	2	2
17	2	2	2
18	2	2	2
19	2	2	2
20	2	2	2
21	2	2	2
22	2	2	2
23	2	2	2
24	2	2	2
25	2	2	2
26	2	2	2
27	2	2	2
28	3	3	3
29	3	3	3
30	3	3	3
31	3	3	3
32	3	3	3
33	3	3	3
34	3	3	3
35	3	3	3
36	3	[2]	[2]
37	3	3	3
38	3	3	3
39	3	3	3
40	3	3	3
41	3	3	3
42	3	3	3
43	3	[2]	[2]
44	3	3	3

Table 4.9: Subject classification HCs=1, PRP=2, SSc=3, (JSS+E, $\alpha_{in} - \alpha_{out}$ 0.15-0.20, subset cardinality 11)

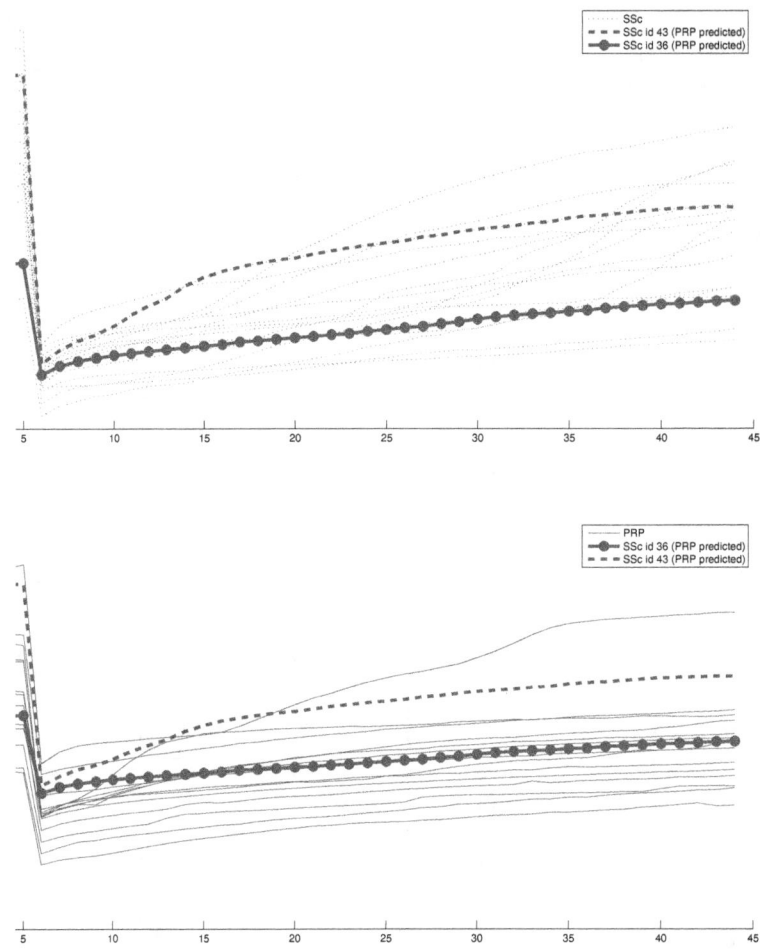

Figure 4.10: Recovery curves for the misclassified subjects (spatial domain) in the original and predicted groups

Moreover, according to the section 2.2, the 3D polar representation in Figure 4.11 shows that the subject (id) 36 has the typical behaviour of the corresponding predicted PRP group (cf Figure 2.5), while the subject (id) 43 shows a typical SSc pattern. In this last case, the misclassification may be due to the connective tissue disease (Sjögren syndrom) which affects the subject.

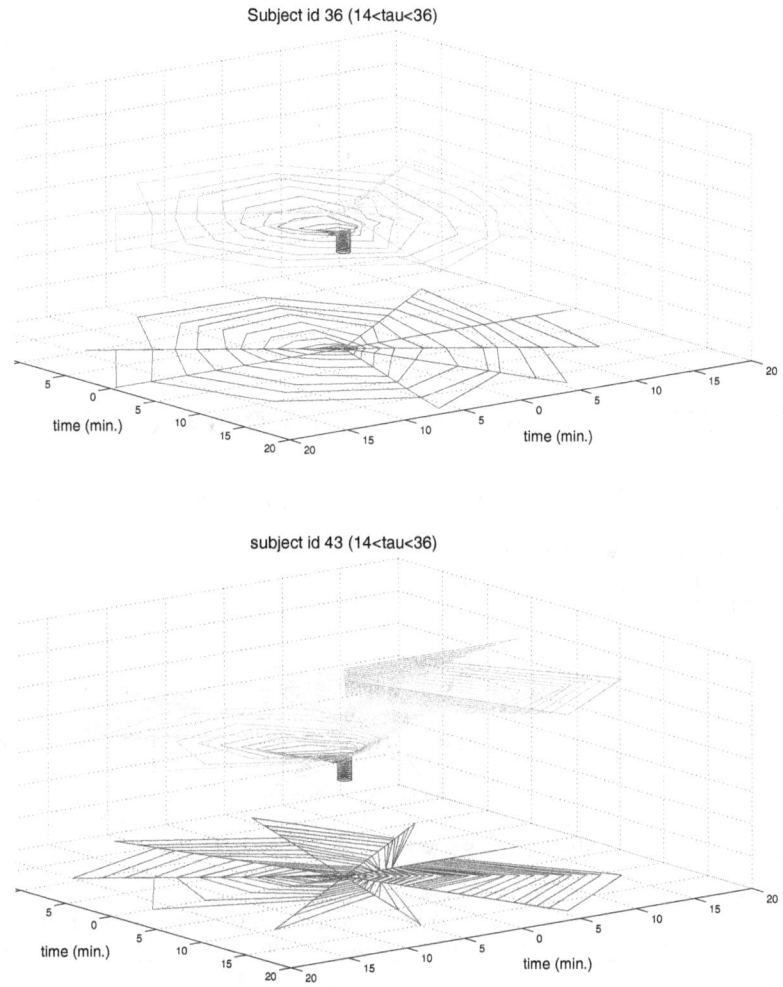

Figure 4.11: 3D Polar representation for the misclassified subjects (spatial domain)

Finally, comparing these results with those from a stepwise selection algorithm (setting the subset cardinality to 11 features and $\alpha_{in} - \alpha_{out}$ to 0.15-0.20), we observe the higher permormance of the JSS+E procedure. In fact, while the confusion matrix for the stepwise procedure in Table 4.10 gives a 4.5% estimate of the apparent error rate, the leave-one-out cross-validation procedure reports an estimated LOCV error rate strongly higher (18, 2%) with respect to the JSS+E procedure (4.5%). For a more detailed discussion about the stepwise selection procedure performance in the spatial domain, see [Aretusi et al., 2010c].

		Classified		
		HCs	PRP	SSc
Training set	HCs	12	0	1
	PRP	0	14	0
	SSc	0	1	16
	APER	0,045		
Cross-validation	HCs	10	1	2
	PRP	0	13	1
	SSc	2	2	13
	LOCVER	0,182		

Table 4.10: Confusion Matrix (stepwise selection, $\alpha_{in} - \alpha_{out}$ 0.20-0.25, subset cardinality 11)

4.2.2 FLDA in the Spatio-Temporal Domain Approach

Enumerating the features from 1 to 279 (F.id) according to the section 3.3.2, in the spatio-temporal domain we choose to select (by the JSS+E procedure) the following subset of variables:

F.id	Feature	Description
18	$\hat{\beta}_{(1,1,0)2}$	GMRF interaction parameter for the diagonal south-west (north-east) direction (left hand)
21	$\hat{\beta}_{(0,1,1)2}$	GMRF interaction parameter for the vertical direction with a temporal lag 1 (left hand)
49	$T_2(2,0,0)$	Homogeneity, horizontal direction lag 2
69	$T_2(6,0,0)$	Homogeneity, horizontal direction lag 6
127	$T_1(6,-6,0)$	Contrast, diagonal south-east (north-west) direction lag 6 with a temporal lag 1
138	$T_4(2,2,0)$	Correlation, diagonal south-west (north-east) direction lag 2
165	$T_3(2,0,1)$	Energy, horizontal direction lag 2 with a temporal lag 1
185	$T_3(6,0,1)$	Energy, horizontal direction lag 6 with a temporal lag 1

Table 4.11: JSS+E algorithm best subset (JSS+E, $\alpha_{in} - \alpha_{out}$ 0.20-0.25, subset cardinality 8)

Specifically, two of the selected variables are related to the parameters of the GMRF related to the vertical and diagonal south-west (north-east) directions, while the remaining ones are represented by the indices T_1, T_2, T_3 and T_4. Variables related to the diagonal directions are characterised by displacements of 2 and 6 lags, while variables related to the horizontal one are defined for spatial lags ranging from 2 to 6. The wider spatial lag observed for the horizontal direction, probably captures information about the orderliness between fingers. Specifically, In basal condition, the cutaneous temperature distribution is largely effected by the thermal properties of the underlying anatomy. Therefore, while the spatial approach selects mostly the vertical direction as the temperature gradient is largely dominated by the thermal track of longitudinal vessels, the response to the thermal stress invokes a larger vascular involvement and a facilitate thermal exchange between vessel and surrounding tissues, thus generating a larger horizontal and diagonal gradient in finger temperature distribution.

Performing an LDA procedure on the selected subset in Table 4.11, we first test the statistical significance of differences between groups means. The following Table 4.12 reports the Wilk's Lambda statistics; the p-values show significance of the test meaning that the differences in the group mean discriminant scores are greater than what could be attributed to the sampling error. The first eigenvalue accounts for a substantial proportion of the total. In fact, the importance of the first root is 0,753; thus the mean vectors lie largely in the first dimension. The proportion of the second eigenvalue on the total is 0,247. The squared canonical correlations between each of the two discriminant functions and the grouping (dummy) variables are 0,744 and 0,489, respectively.

Eigenvalues	C.V. importance	Roy's statistic	Wilk's Lambda	F	P-value
2,910	0,753	0,744	0,131	7,504	0,000
0,956	0,247	0,489	0,511	4,782	0,001

Table 4.12: Wilk's Lambda (JSS+E, $\alpha_{in} - \alpha_{out}$ 0.20-0.25, subset cardinality 8)

The Figure 4.12 on the following page depicts the canonical variables produced by the model; the subjects seems to clusterise in 3 groups.

The confusion matrix resulting from LDA gives a 9% estimate both of the apparent error rate and the LOCV error rate computed with the cross validation procedure. Specifically, Table 4.13 on page 75 shows the confusion matrix computed for the training set and for the leave-one-out cross-validation procedure. It can be seen that, 2 HCS, 1 PRP and 1 SSc are wrongly classified.

These correspond to the individuals (id) 1, 9, 26 and 36 (cf Table 4.15) for which we have additional clinical informations. We note that:

- the individuals 1 and 9 are heavy smokers. Heavy smokers tend to maintain an higher than normal vasoconstrictor tone and a delayed response to external stimuli which

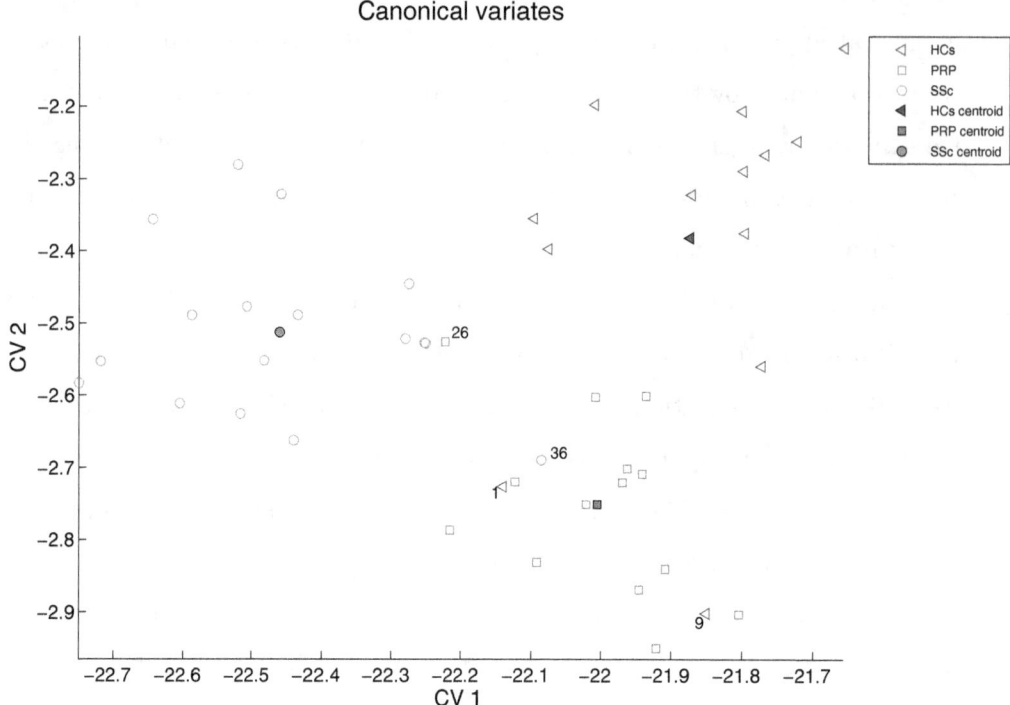

Figure 4.12: Canonical variates (JSS+E, $\alpha_{in} - \alpha_{out}$ 0.20-0.25, subset cardinality 8)

request vasodilatation processes. For this reason, while their steady state temperature distribution allows right classification, their dynamic response closely resembles idiopathic Raynauds Phenomenon behaviour;

- subject (id) 26 suffers for bilateral carpal trauma. The presence of inflammatory processes alters normal thermoregulatory processes, both at basal and under functional stimulation. In particular, as the involvement in this patient is not symmetrical between wrists, it is possible that he/she shows like SSc behaviour as the nerve involvement differently effects each single finger;

- subject (id) 36 is a longstanding diffuse SSc for which the vascular damage is well advanced. In this condition, the cutaneous thermoregulation is systematically damaged and no vaso-active reaction can be expressed. Therefore, only passive thermal exchange with the surrounding warmer environment can help into re-establishing basal

		Classified		
		HCs	PRP	SSc
Training set	HCs	11	2	0
	PRP	0	13	1
	SSc	0	1	16
	APER	0,090		
Cross-validation	HCs	11	2	0
	PRP	0	13	1
	SSc	0	1	16
	LOCVER	0,090		

Table 4.13: Confusion Matrix (JSS+E, $\alpha_{in} - \alpha_{out}$ 0.20-0.25, subset cardinality 8)

condition, as the PRP case. By the way, in this case, early assessment through functional thermal test is not significant as the diagnosis is extremely well assessed by clinic features.

Finally, we remark that subject (id) 43 suffers for Sjögren syndrome for which no studies about the fingers thermoregulation process are available to establish possible differencies with SSc or PRP features. However, this syndrome resembles closely both clinical, autoimmune, haematic features of SSc, with particular reference to the vasomotor tone and vascular impairment. Therefore the functional test provides responses very close to SSc that cannot be detected in the basal condition (spatial-domain). This might explain the misclassifications.

Table 4.14 shows the estimated standardised coefficients of the discriminant functions which provide information on the relative impact of the predictors on the dependent variable. We observe that the variables with F.id 185, 165, 69 contribute most to separating the groups

	Functions	
Feature id	1	2
18	-0,744	0,342
21	1,013	0,100
49	-3,864	-0,611
69	3,294	-1,487
127	-1,482	-0,686
138	-0,874	0,336
165	-2,772	-2,724
185	2,309	3,604

Table 4.14: Standardised coefficients for the two discriminant functions (JSS+E, $\alpha_{in} - \alpha_{out}$ 0.15-0.20, subset cardinality 8)

in the two discriminant functions. Furthermore, the variables 49 and 127 have a certain importance with respect to the first function and the variable 69 contribute to the second function.

The following Table 4.15 lists the subjects (id) and the corresponding group prediction both considering the training set and performing a leave-one-out cross-validation procedure.

To get more insight in the misclassifications, we plot (Figure 4.13 on page 78) the re-warming curves (cf Section 2.1) of the misclassified subjects both respect to the orignal and predicted group that seem to indicate a different clusterization for the misclassified subjects and to confirm the predicted groups. Moreover, we use the 3D polar representations (cf section 2.2); we note that the subject (id) 36 in Figure 4.11 on page 71 seems a typical secondary RP. Also, the Figure 4.14 on page 79 shows that the other misclassified subjects have the typical patterns (cf Figure 2.5 on page 15) of the corresponding predicted groups.

Finally, comparing these results with those from a stepwise selection algorithm, we observe the higher permormance of the JSS+E procedure. The stepwise procedure selects 7 features when setting the maximal subset cardinality to 8 features and $\alpha_{in} - \alpha_{out}$ to (0.15-0.20). The Table 4.16 represents the confusion matrix for the stepwise procedure and gives an $18,18\%$ estimate of the apparent error rate and an estimated LOCV error rate of $18,18\%$, strongly higher respect to the JSS+E procedure (9%). For a more detailed discussion about the stepwise selection procedure performance in the spatio-temporal doamin, see [Aretusi et al., 2010c].

id	Group	Classified	Cross-validate
1	1	[2]	[2]
2	1	1	1
3	1	1	1
4	1	1	1
5	1	1	1
6	1	1	1
7	1	1	1
8	1	1	1
9	1	[2]	[2]
10	1	1	1
11	1	1	1
12	1	1	1
13	1	1	1
14	2	2	2
15	2	2	2
16	2	2	2
17	2	2	2
18	2	2	2
19	2	2	2
20	2	2	2
21	2	2	2
22	2	2	2
23	2	2	2
24	2	2	2
25	2	2	2
26	2	[3]	[3]
27	2	2	2
28	3	3	3
29	3	3	3
30	3	3	3
31	3	3	3
32	3	3	3
33	3	3	3
34	3	3	3
35	3	3	3
36	3	[2]	[2]
37	3	3	3
38	3	3	3
39	3	3	3
40	3	3	3
41	3	3	3
42	3	3	3
43	3	3	3
44	3	3	3

Table 4.15: Subject classification HCs=1, PRP=2, SSc=3, (JSS+E, $\alpha_{in} - \alpha_{out}$ 0.20-0.25, subset cardinality 8)

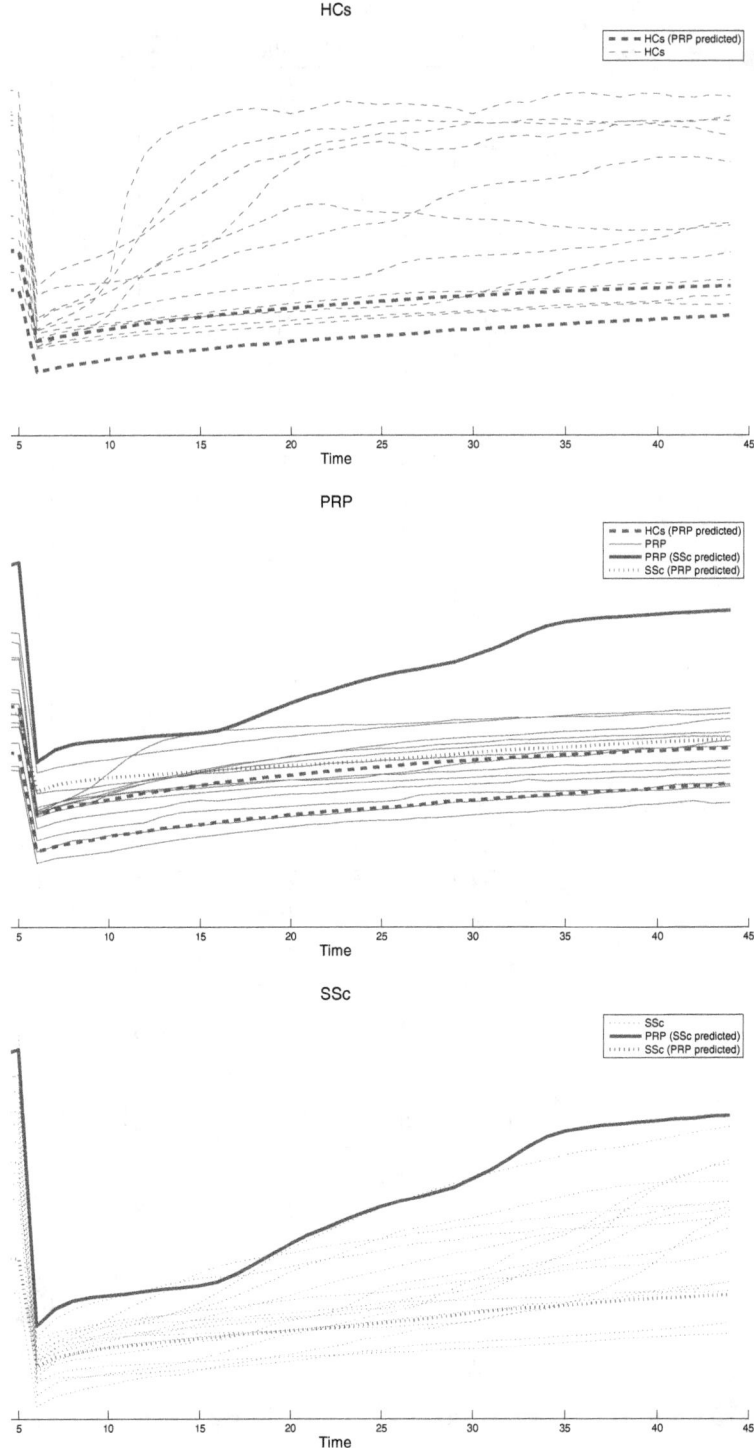

Figure 4.13: Recovery curves for the misclassified subjects (spatio-temporal domain) in the original and predicted groups

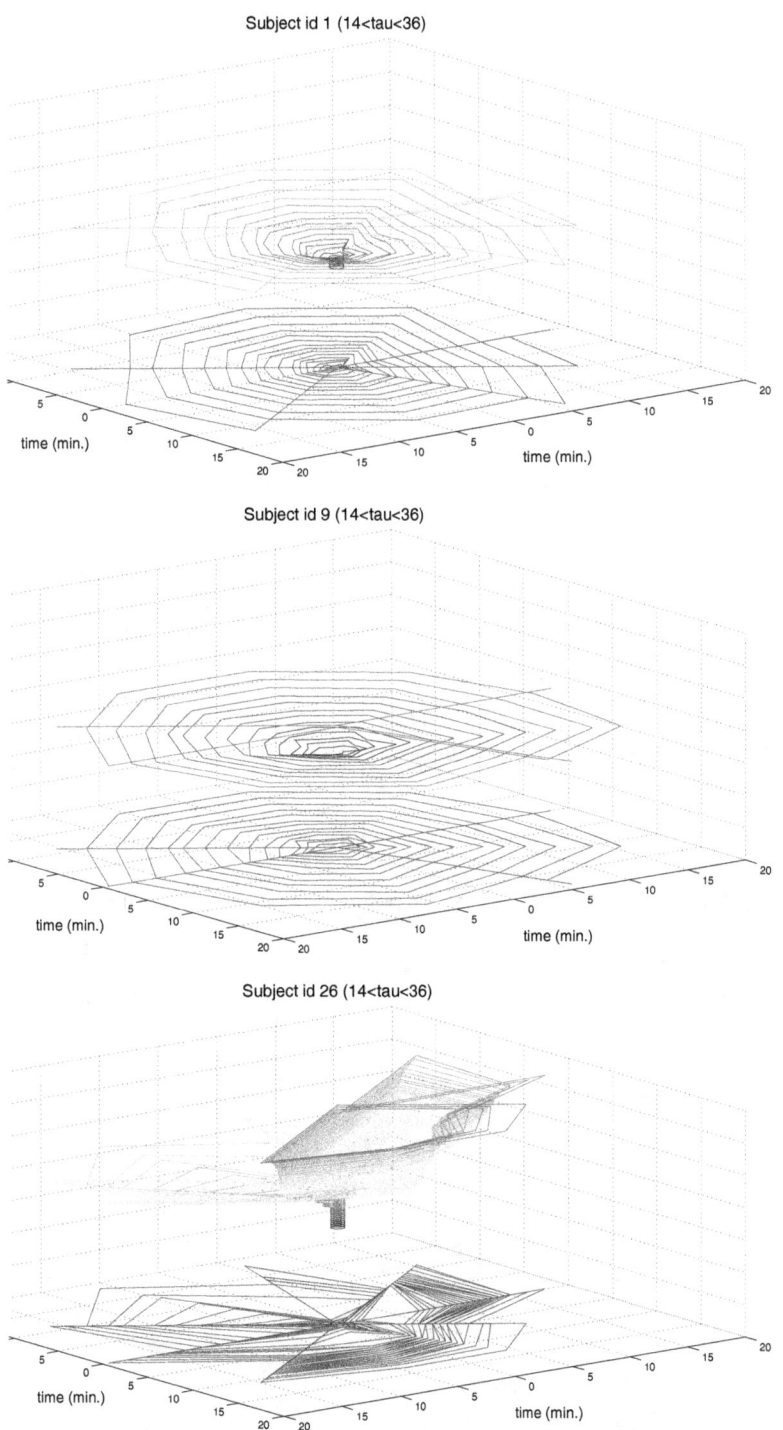

Figure 4.14: 3D Polar representation for the misclassified subjects (spatio-temporal domain)

		Classified		
		HCs	PRP	SSc
Training set	HCs	10	3	0
	PRP	0	12	2
	SSc	1	2	14
	APER	0,182		
Cross-validation	HCs	10	3	0
	PRP	0	12	2
	SSc	1	2	14
	LOCVER	0,182		

Table 4.16: Confusion Matrix (Stepwise selection, $\alpha_{in} - \alpha_{out}$ 0.20-0.25, maximal subset cardinality 8)

Conclusions

In this study we have discussed the problem of classifying infrared imaging data for differential diagnosis of primary and secondary RP forms. The segmentation and registration of IR images were also considered. Specifically, we have noticed that segmentation of IR images is not an easy task. It is true that the infrared image acquisition is a part of the problem in the infrared segmentation process. In fact, the technologies required for IR imaging are much less mature than the ones used in visible imaging. However, nowadays, cameras that collect simultaneously both visual and thermal images are available and performing a fusion of the two types of images may lead to an improvement of the results.

Classification of primary and secondary RP forms have been performed both in a spatial and in a spatio-temporal context using texture analysis techniques. Since the images are not stationary in time and in space (Figure 3.9), a necessary first step was that of studying the spatial and spatio-temporal trend and correlation structure of the heating distribution (large and small scale variability).

Furthermore, the relevant features for the supervised classification of the RP are unknown *a priori*. Thus, we have proposed two different techniques which were able to generate a large number of discriminant variables. Due to the characteristics of the generated data set, of primary importance are of course those classifiers that can provide a rapid analysis of large amounts of data and have the ability to handle *fat data*, where there are more variables than observations. Therefore, besides considering the stepwise subset selection procedure, we propose a new subset selection algorithm, called JSS+E (Jackknifed Stepwise Selection with Exhaustive search) in order to reduce the bias of the misclassification error rate estimation and to improve the results of the classification with respect to the stepwise selection procedure.

Good classification results, characterised by an error rate of 9% in the spatio-temporal domain and of 4,5% in the spatial domain, were achieved through FLDA; to our knowledge there are no other works which have achieved such results for Raynaud's Phenomenon observed through the protocol described in the section 1.3. In particular, the spatial framework is extremely interesting since, in this case, it would be possible to recalibrate the experimental protocol avoiding the cold stress test. This would have enormous advantages in order to perform a completely non-invasive experiment and to reduce costs and time of computing and data collection. Furthermore, very often the subjects affected by the RP (or other related diseases) are not able to completely perform the experiment, in particular the cold stress test. As an example, the person in Figure 4.15 is affected by a rheumatoid arthritis in an advanced stage. Thus, this subject is not able to place the hands in the correct position for a long time, which make it impossible to perform the experiment as required by the protocol described in section 1.3.

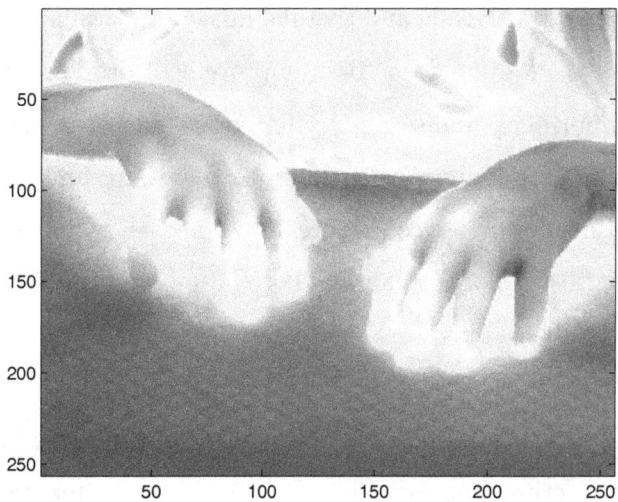

Figure 4.15: Example of a subject affected by a rheumatoid arthritis

Other classifiers, such as Partial Least Square [Barker and Rayens, 2003] or shrinkage methods (e.g., DALASS) [Trendafilov and Jolliffe, 2007] could be considered and this will be of particular interest for future works.

Moreover, measuring the strength of the infrared signal in certain frequency bands, can also be helpful to classify the RP [Aretusi et al., 2009]. In fact, the spectral density is a useful technique in understanding the underlying structure or any salient features in a stochastic process; it allows a valuable insight into the probabilistic structure of the process since the spectral density function is a decomposition of the total power contained in the signal. As an example, the following Figure 4.16 shows the profile of the bidimensional spatial spectra estimated along the diagonal for our RP patients sample.

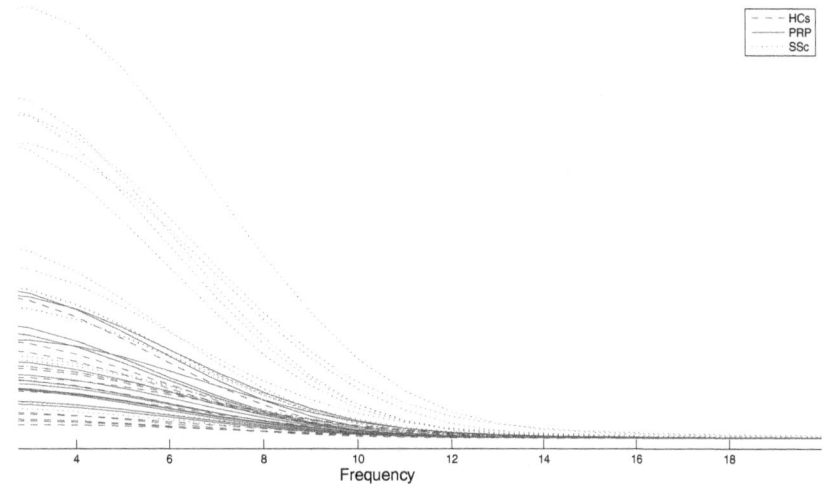

Figure 4.16: Diagonal of the (bidimensional) spatial spectra estimated in the frequency band [1, 18]

We observe that the spectra seems to differentiate between HCs, PRP and SSc. If we also consider Figure 4.17, which represents the group mean power logspectra in certain frequency band (e.g., the band [1, 18]), the differences between the groups seem even more clear.

As final considerations, we do recognise that the small sample size of the study could be an obstacle for having a strong interpretation of the statistical results. Unfortunately, the small sample size is something that can hardly be changed in practical problems like this but we aim at extending the analysis to a wider data set (presently being acquired) in further works.

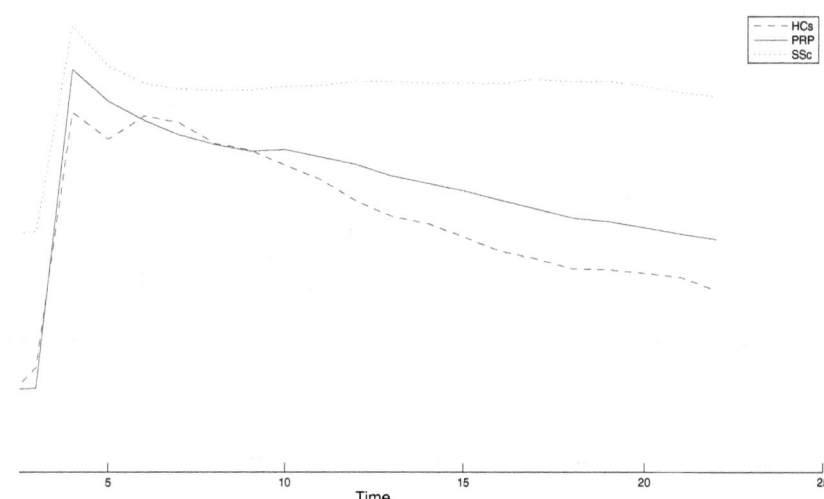

Figure 4.17: Group mean power logspectra in the frequency band $[1, 18]$

Bibliography

H. Almuallim and T.G. Dietterich. Learning boolean concepts in the presence of many irrilevant features. *Artifcial Intelligence*, 69:279–305, 1994.

K. Ammer. Diagnosis of Raynaud's Phenomenon by Thermography. *Skin research and technology*, 2:182–185, 1996.

J.A. Anderson. *Logistic Discrimination*, volume 2 of *Handbook of Statistics: Classification Pattern Recognition and Reduction Dimensionality*, pages 169–191. P.R. Krishnaiah and L.N. Kanal (Eds), 1982.

M.E. Anderson, T.L. Moore, M. Lunt, and A.L. Herrick. The Distal-Dorsal Difference: a Thermographic Parameter by which Differentiate Between Primary and Secondary Raynaud's Phenomenon. *Rheumatology*, 46:533–538, 2007.

G. Aretusi, L. Ippoliti, A. Merla, and T. Subba Rao. Spatial Spectral Method for Diagnosis of Raynaud's Phenomenon Based on Functional Infrared Imaging. Technical report available from the authors upon request, 2009.

G. Aretusi, L. Fontanella, L. Ippoliti, and A. Merla. Space-Time Texture Analysis in Thermal Infrared Imaging for Classification of Raynaud's Phenomenon. *Complex data modelling and computationally intensive statistical methods*, Springer, ISBN: 978-88-470-1385-8, 2010a.

G. Aretusi, L. Fontanella, L. Ippoliti, and A. Merla. Supervised Classification of Thermal High-Resolution IR Images for the Diagnosis of Raynaud's Phenomenon. *New Perspectives in Statistical Modelling and Data Analysis*, Springer, ISBN: 978-3-642-11362-8, 2010b.

G. Aretusi, C.C. Taylor, L. Ippoliti, and A. Merla. Spatio-Temporal Functional Infrared Imaging Analysis for Classification of Raynaud's Phenomenon. Technical report submitted for publication and available from the authors upon request, 2010c.

M. Barker and W.N.T. Rayens. Partial least squares for discrimination. *Journal of Chemometrics*, 17:166–173, 2003.

J. Belch. Raynaud's phenomenon. its relevance to scleroderma. *Ann. Rheum. Dis.*, (50): 839–845, 2005.

J.E. Besag and A.P. Moran. On the estimation and testing of spatial interaction in gaussian lattice processes. *Biometrika*, 65:555–562, 1975.

J.A. Block and W. Sequeira. Raynaud's phenomenon. *Lancet*, (357):2042–2048, 2001.

L.G. Brown. A survey of image registration techniques. *ACM Computing Surveys*, (24): 326–376, 1992.

J.S. Chang, H.Y.M. Liao, M.K. Hor, J.W. Hsieh, and M.Y. Cgern. New automatic multi-level thresholding technique for segmentation of thermal images. *Images and vision computing*, (15):23–34, 1997.

S. Clark, F. Campbell, T. Moore, M.I.V. Jayson, T.A. King, and A.L. Herrick. Laser doppler imaging. a new technique for quantifying microcirculatory flow in patients with primary raynaud's phenomenon and systemic sclerosis. *Microvasc. Res.*, 57:284–291, 1999.

J.P. Cocquerez and S. Philipp. *Analyse d'images : filtrage et segmentation*. Masson, Paris, 1995.

T.M. Cover and J.A. Thomas. *Elements of Information Theory*. John Wiley & Sons, 1991.

N.A. Cressie. *Statistics for Spatial Data*. John Wiley & Sons, New York, 2nd edition, 1993.

M. Dash and H. Liu. Feature selection for classification. *Intelligent Data Analysis*, 1: 131–156, 1997.

A. Di Carlo. Thermography and the possibilities for its applications in clinical and experimental dermatology. *Clin. Dermatol.*, 13:329–336, 1995.

S. Di Zio, L. Ippoliti, and A. Merla. Supervised Classification of Functional Infrared Imaging Data for Differential Diagnosis of Raynaud's Phenomenon. Technical report submitted for publication and available from the authors upon request, 2009.

I. Dryden. General shape and registration analysis. In *In*. Chapman and Hall, 1997.

I. Dryden, L. Ippoliti, and L. Romagnoli. Adjusted maximum likelihood and pseudo-likelihood estimation for noisy gaussian markov random fields. *J Comput Graph Stat*, 11:370–388, 2002.

B. Efron. Estimating the error rate of a prediction rule: improvement on cross-validation. *Journal of the American Statistical Association*, 78:316–331, 1983.

R.A. Fisher. The Use of Multiple Measurements in Taxonomic Problems. *Ann. Eugenics*, 7:179–188, 1936.

L. Fontanella, L. Ippoliti, R.J. Martin, and S. Trivisonno. Interpolation of spatial and spatio-temporal gaussian fields using gaussian markov random fields. *Advanced in Data Analysis and Classification*, 2:63–79, 2008.

K. Fukunaga. *Introduction to Statistical Pattern Recognition*. Academic Press, New York, 2nd edition, 1990.

R.C. Gonzalez, R.E. Woods, and S.L. Eddins. *Digital Image Processing Using Matlab*. Pearson Education Ltd, London, 2004.

P.J. Green and B.W. Silverman. *Nonparametric Regression and Generalized Linear Models: A Roughness Penalty Approach*. Chapman and Hall, London, 1994.

J. Gullberg. *Mathematics from the Birth of Numbers*. W.W. Norton and Company, New York, 1997.

M. Hahn, C. Hahn, M. Jünger, A. Steins, D. Zuder, T. Klyscz, A. Bchtemann, G. Rassner, and V. Blazek. Local cold exposure test with a new arterial photoplethysmography sensor in healthy controls and patients with secondary raynaud's phenomenon. *Microvasc. Res.*, 57:187–198, 1999.

R.M. Haralick, K. Shanmugam, and I. Dinstein. Textural features for image classification. *IEEE Trasnaction on Systems, Man, and Cybernetics*, SMC-3(6):610–621, 1973.

T. Hastie, R. Tibshirani, and J. Friedman. *The Elements of Statistical Learning: Data Mining, Inference, and Prediction*. Springer, 2nd edition, 2009.

D. Haussler and M. Opper. Mutual information, metric entropy and cumulative relative entropy risk. *The Annals of Statistics*, 25(6):2451–2492, 1997.

D.M. Hawkins. The Subset Problem in Multivariate Analysis of Variance. *JRSS*, 38(2): 132–139, 1976.

R. Heriansyak and S.A.R. Abu-Bakar. Defect detection in thermal image for nondestructive evaluation of petrochemical equipments. *NDT&E International*, (42):729–740, 2009.

A.L. Herrick and S. Clark. Quantifying digital vascular disease in patients with primary raynaud's phenomenon and systemic sclerosis. *Ann. Rheum. Dis.*, 57:70–78, 1998.

C. Ibarra-Castanedo, D. Gonzales, M. Klein, M. Pilla, S. Vallerand, and X. Maldague. Infrared image processing and data analysis. *Infrared Physics & Technology*, (46):75–83, 2004.

A. Jarc, J. Pers, P. Rogelj, M. Perse, and S. Kovacic. Texture features for affine registration of thermal and visible images. Computer Vision Winter Workshop, 2007.

R.A. Johnson and D.W. Wichern. *Applied Multivariate Statistical Analysis*. Pearson education. Prentice Hall, London, 6th edition, 2007.

J.T. Kent and K.V. Mardia. *The link between kriging and thin-plate splines*, pages 324–339. Probabilty, Statistics and Optimization. Wiley, New York, 1994.

W.J. Krzanowsky. *Principles of Multivariate Analysis*. Oxford University Press, Oxford, revised ed. edition, 2003.

W.J. Krzanowsky and F.H.C. Marriott. *Multivariate Analysis: Classification, Covariance Structures and Repeated Measurements*, volume 2. Arnold (Holder Headline Group), London, 1979.

A. Kuryliszin-Moskal. Soluble adhesion molecules (svcam-1, se-selectin), vascular endothelial growth factor (vegf) and endothelin-1 in patient with systemic sclerosis: relationship to organ systemic involvement. *Clin. Rheumatol.*, (24):111–116, 2005.

G.H. Landeweerd and E.S. Gelsema. The Use of Nuclear Texture Parameters in the Automatic Analysis of Leukocytes. *Pattern Recognition*, 10:57–61, 1978.

X.P.V. Maldague. *Theory and practice of infrared technology for nondestructive Testing*. Wiley Interscience, 2001.

K.V. Mardia, J.T. Kent, and J.M. Bibby. *Multivariate Analysis*. Academic Press, Duluth, London, 1979.

K.V. Mardia, J.T. Kent, C.R. Goodall, and J.A. Little. Kriging and splines with derivative information. *Biometrika*, 83(1):207–221, 1996.

G.J. McLachlan. *Discriminant Analysis and Statistical Pattern Recognition*. John Wiley & Sons, Hoboken, New Jersey, 2004.

A. Merla, G.L. Romani, S. Di Luzio, L. Di Donato, G. Farina, M. Proietti, S. Pisarri, and F. Salsano. Infrared functional imaging applied to raynaud's phenomenon: infrared functional imaging applied to diagnosis and drug effect. *Int. J. Immunopathol. Pharmacol.*, 15(1):41–52, 2002a.

A. Merla, Di L. Donato, S. Pisarri, M. Proietti, F. Salsano, and G.L. Romani. Infrared functional imaging applied to raynaud's phenomenon. *IEEE Eng. Med. Biol. Mag.*, 21 (6):73–79, 2002b.

A. Miller. *Subset Selection in Regression*. Chapman and Hall/CRC, London, 2nd edition, 2002.

R.G. Miller. The jackknife - a review. *Biometrika*, 61:1–15, 1974.

A.V. Moore and M.S. Lee. Efficient algorithms for minimizing cross validation error. In *Proceedings of the Eleventh International Conference on Machine Learning*, pages 190–198, New Brunswick, New Jersy, 1994. Morgan Kaufmann.

D. O'Reilly, L. Taylor, K. El-Hadivi, and M.I. Jayson. Measurement of cold challenge response in primary Raynaud's phenomenon and Raynaud's phenomenon associated with systemic sclerosis. *Ann. Rheum. Dis.*, 51(11):1193–1196, 1992.

W.K. Pratt. *Digital Image Processing*. Wiley Interscience, 4th edition, 2007.

M.H Quenouille. Approximate tests of correlation in time series. *Journal of the Royal Statistical Society*, 11:18–44, 1949.

M.H Quenouille. Notes on bias in estimation. *Biometrika*, 61:353–360, 1956.

J.O. Ramsay and B.W. Silverman. *Functional Data Analysis*. Springer-Verlag, New York, 1997.

A.C. Rencher. *Methods of Multivariate Analysis*. Wiley-Interscience, 2nd edition, 2002.

S. Sato, M. Hasegawa, K. Takehara, and T.F. Tedder. Altered b lymphocyte function induces systemic autoimmunity in systemic sclerosis. *Mol. Immunol.*, (42):821–831, 2005.

O. Schuhfried, G. Vacariu, T. Lang, M. Korpan, H.P. Kiener, and V. Fialka-Moser. Thermographic Parameters in the Diagnosis of Secondary Raynaud's Phenomenon. *Arch. Phys. Med. Rehabil.*, 81:495–499, 2000.

D.A. Scribner, J.M. Schuller, P. Warren, J.G. Howard, and M.R. Kruer. Image preprocessing for the infrared. *Proceedings of SPIE, the International Society for Optical Engineering*, (4028):222–233, 2000.

J.L. Semmlow. *Biosignal and Biomedical Image Processing*. CRC Press, 2004.

B. Sing-Tze. *Pattern recognition and image preprocessing*, volume 14 of *Signal processing and communications*. CRC Press, 2nd edition, 2002.

E.W. Steyerberg, M.J.C. Eijkemans, F.E. Harrel Jr, and J.D.F. Habbema. Prognostic modelling with logistic regression analysis: a comparison of selection and estimation methods in small data sets. *Statistics in medicine*, 19:1059–1079, 2000.

M. Stone. Cross-validatory choice and assessment of statistical predictions (with discussion). *Journal of the Royal Statistical Society*, 36:3–23, 1974.

R. Sutton and E.L. Hall. Texture Measurse for Automatic Classification of Pulmonary Disease. *IEEE Transactions on Computers*, C-21:667–676, 1972.

B. Thompson. Stepwise Regression and Stepwise Discriminant Analysis Need Not Apply here: A Guidelines Editorial. *Educational and Psychological Measurement*, 55(4):525–534, 1995.

B. Thompson. Significance, Effect Sizes, Stepwise Methods, and Other Issues: Strong Arguments Move the Field. *The journal of Experimental Education*, 70(1):80–93, 2001.

N.T. Trendafilov and I.T. Jolliffe. Dalass: Variable selectoin in discriminant analysis via the lasso. *Computational Statistics & Data Analysis*, 51:3718–3736, 2007.

G. Wahba. *Spline Models for Observational Data*. regional conference series in applied mathematics. Society for Industrial and Applied Mathematics, Philadelphia, 1990.

A.P. White and W.Z. Liu. The jack-knife with a stepwise discriminant algorithm - a warning to BMDP users. *Journal of Applied Statistics*, 20(1):187–190, 1993.

B. Zitova and J. Flusser. Image registration methods: a survey. *Image and Vision Computing*, (21):977–1000, 2003.

www.ingramcontent.com/pod-product-compliance
Lightning Source LLC
Chambersburg PA
CBHW081048170526
45158CB00006B/1902